Studies in Fuzziness and Soft Computing

Volume 346

Series editor

Janusz Kacprzyk, Polish Academy of Sciences, Warsaw, Poland
e-mail: kacprzyk@ibspan.waw.pl

About this Series

The series "Studies in Fuzziness and Soft Computing" contains publications on various topics in the area of soft computing, which include fuzzy sets, rough sets, neural networks, evolutionary computation, probabilistic and evidential reasoning, multi-valued logic, and related fields. The publications within "Studies in Fuzziness and Soft Computing" are primarily monographs and edited volumes. They cover significant recent developments in the field, both of a foundational and applicable character. An important feature of the series is its short publication time and world-wide distribution. This permits a rapid and broad dissemination of research results.

More information about this series at http://www.springer.com/series/2941

Alexander Vitalievich Bozhenyuk
Evgeniya Michailovna Gerasimenko
Janusz Kacprzyk
Igor Naymovich Rozenberg

Flows in Networks
Under Fuzzy Conditions

Springer

Alexander Vitalievich Bozhenyuk
Southern Federal University
Taganrog
Russia

Evgeniya Michailovna Gerasimenko
Southern Federal University
Taganrog
Russia

Janusz Kacprzyk
Systems Research Institute
Polish Academy of Sciences
Warsaw
Poland

Igor Naymovich Rozenberg
Public Corporation "Research
    and Development Institute of Railway
    Engineers"
Moscow
Russia

ISSN 1434-9922   ISSN 1860-0808   (electronic)
Studies in Fuzziness and Soft Computing
ISBN 978-3-319-82398-0   ISBN 978-3-319-41618-2   (eBook)
DOI 10.1007/978-3-319-41618-2

Printed on acid-free paper

This Springer imprint is published by Springer Nature
The registered company is Springer International Publishing AG Switzerland

Contents

Introduction

Urgency of the flow problems including the maximum flow finding, the minimum cost flow finding, the maximum dynamic flow finding, and the minimum cost dynamic flow finding lies in the fact that economic development of any country is caused by the presence of roads. Due to the process of urbanization, the number of vehicles has been increased and the quality of roads is poor; there is no an adequate policy regarding the construction of the new roads and repair existing ones. Important sphere of researches is the flow control. The expanding transport infrastructure leads to congestion, emergencies due to "traffic jams."

Scientists develop and solve various optimization problems in transportation networks, in particular the tasks of identifying the congested roads on the map and finding the routes of the minimum cost. Dynamic networks are important subject of research, as they allow solving flow problems, taking into account the time factor, thereby one can find the optimal routes of the cargo transmission considering the time factor, to find the maximum cargo traffic between selected points on the road map within a specified time period.

The complexity of the factors influencing the network parameters, in particular their inherent uncertainty, is not taken into account in the flows literature. In fact, the flow bounds, the values of flows passing along the arcs, and transmission costs cannot be accurately measured according to their nature. The weather conditions, traffic jams, and repairs influence flow bounds. Variations in petrol prices can influence transmission costs. Therefore, the flow tasks in static and dynamic networks will be proposed in the conditions of incomplete and inaccurate information.

The present book deals with the problems of finding the flows in static and dynamic networks in terms of fuzziness and partial uncertainty. The solution of these problems is based on the concepts of fuzzy logic presented by the authors Bellman and Zadeh [1], Zimmermann [2], Berstein and Korovin [3], Borisov [4], Dubois and Prade [5], and others.

Methods for solving flow problems in transportation networks were considered by Ford and Fulkerson [6], Minieka [7], Busacker and Gowen [8], Christofides [9], and others. In the fuzzy conditions, these problems were considered by the authors S. Chanas and W. Kolodziejczyk [10], Fonoberova and Lozovanu [11] and others.

However, these studies do not take into account the existence of nonzero lower arc flow bounds, reflecting the transportation profitability degree and dependence of the fuzzy network parameters from the departure time.

The methods of the maximum and the minimum cost flow finding with fuzzy nonzero lower, upper flow bounds and transmission costs; the maximum and the minimum cost dynamic flow finding with fuzzy transit parameters, such as nonzero lower, upper flow bounds and transmission costs, are proposed in the present book.

References

1. Bellman RE, Zadeh LA (1970) Decision making in a fuzzy environment. Manag Sci 17:141–164
2. Zimmermann HJ (1991) Fuzzy set theory and its applications, 2nd edn. Kluwer Academic Publishers, Boston/Dordrecht/London
3. Berstein LS, Korovin SYa, Melikhov AN (1985) Compressing a set of reference situations in linguistic models of situational control . Avtomat i Telemekh No. 2:118–123
4. Borisov AN, Alekseev AV, Merkuryeva GV, Slyadz NN, Glushkov VI (1989) Fuzzy information processing in decision-making systems. Radio and Communication Publisher, Moscow
5. Dubois D, Prade H (1978) Operations on fuzzy numbers. Int J Syst Sci 9(6):613–626
6. Ford LR, Fulkerson DR (1962) Flows in networks. Princeton University Press, Princeton
7. Minieka E (1978) Optimization algorithms for networks and graphs. Marcel Dekker, Inc., New York and Basel
8. Busacker RG, Gowen P (1961) A procedure for determining a family of minimum-cost network flow patterns. Technical Report 15, Operations Research Office, John Hopkins University
9. Christofides N (1975) Graph theory: an algorithmic approach. Academic Press, New York, London, San Francisco
10. Chanas S, Kolodziejczyk W (1982) Maximum flow in a network with fuzzy arc capacities/ S. Chanas. Fuzzy Sets Syst 8(2):165–153
11. Fonoberova MA, Lozovanu DD (2004) The maximum flow in dynamic networks. Comp Sci J Moldova 12, No. 3(36), 387–396

Chapter 1
Flow Tasks in Networks in Crisp Conditions

1.1 Basic Concepts of the Flow Theory

The flow tasks arising in the study of transportation networks are relevant due to their wide practical application. Consider the concept of the transportation network submitted in Definition 1.1.

Definition 1.1 Transportation network (network) is a finite connected directed graph $G = (X, A)$ without loops, where X—the set of nodes, representing the set of the crossroads or stations, A—the set of arcs, i.e., sections of roads connecting the nodes $x_i, x_j \in X$, where $(i, j = 1, \ldots, n)$. Some nonnegative number u_{ij} called arc capacity is set to each arc $(x_i, x_j) \in A$ and there are two nodes: the source $x_0 = s$, which has no incoming arcs, and the sink $x_n = t$, which has no exiting arcs [1].

The parameter c_{ij} determining the transmission cost of the one flow unit along the arc (x_i, x_j) is defined in the tasks of the minimum cost flow finding.

Consider the concept of the arc capacity provided in the Definition 1.2.

Definition 1.2 The capacity u_{ij} of the arc (x_i, x_j) of the network determines the maximum flow value that can go along the arc (x_i, x_j) [2].

Consider the concept of the flow passing along the arcs of the network as defined in 1.3.

Definition 1.3 The set of numbers ξ_{ij} given on the arcs $(x_i, x_j) \in A$ of the network is called flows, if the following conditions are true according to [2]:

$$\sum_{x_j \in \Gamma(x_i)} \xi_{ij} = \sum_{x_k \in \Gamma^{-1}(x_i)} \xi_{ki} = \begin{cases} v, & x_i = s, \\ -v, & x_i = t, \\ 0, & x_i \neq s, t, \end{cases} \quad (1.1)$$

© Springer International Publishing Switzerland 2017
A.V. Bozhenyuk et al., *Flows in Networks Under Fuzzy Conditions*,
Studies in Fuzziness and Soft Computing 346, DOI 10.1007/978-3-319-41618-2_1

and

$$\xi_{ij} \leq u_{ij}, \quad \forall\, (x_i, x_j) \in A. \tag{1.2}$$

The Eq. (1.1) is the condition of the flow conservation reflecting, that the amount of flow entering any node x_i except the source and the sink is equal to the flow leaving the node x_i except the source and the sink. Restrictions on the arc flows are mentioned in Eq. (1.2).

1.2 Selecting of the Arc Capacities

Thus, arc capacities are the parameters that limit the flows passing along the arcs of the network. Road capacity defines the maximum number of vehicles that can pass along the considered road at a unit of time. In fact, capacities are fundamental parameter in determining the maximum flow. Therefore it is necessary to turn to the procedures of capacities determining [3].

Capacities depend on many factors: the road conditions (the width of the roadway, longitudinal slope, curve radii, visibility distance, etc.), the flow of vehicles, the availability of control resources, climatic conditions, the ability to maneuver vehicles across the width of the roadway, psycho-physiological characteristics of drivers and vehicle design. Changing these factors leads to significant fluctuations in capacities during the day, month, season and year. There are significant variations in the speed, resulting in the large number of vehicles moving in groups, as well as reducing the average speed of the flow with the frequent location of noises on the road [4].

Capacities are considered to divide into theoretical, practical and computational. Theoretical capacity u_{theor} is capacity of the road or its part in terms of constant intervals between cars, homogeneous composition of vehicles (e.g., if the composition contains only passenger cars). If we consider a highway, the theoretical capacity of its strip is 2900 passenger cars per hour.

Practical capacity u_{pract} is a parameter provided in real traffic conditions. There are two types of it: the maximal practical u_{max} and practical u_{pract} in real driving conditions. Maximal practical capacity is capacity of the part of road under reference conditions. Practical capacity in specific road conditions corresponds to the areas of roads with the worst road conditions in compared with the reference site [4].

The computational flow capacity u_{comp} determines economically reasonable number of vehicles that can pass along the road (its area) in the specific road conditions and the particular organization of movement. Computational flow capacity can be calculated by Eq. (1.3):

$$u_{comp} = k_n u_{theor} \tag{1.3}$$

In (1.3) k_n—the transition coefficient from the theoretical capacity to the computational, which is determined depending on the type of road, road category and terrain.

In real capacities calculations the formula (1.4) for calculating of practical capacity in real traffic conditions can be used:

$$u_{pract} = Bu_{max} \tag{1.4}$$

In (1.4) u_{max} is a constant depending on the number of lanes of the road. The coefficient B is the final reduction factor of capacity, equals to the product of partial factors $\beta_1 \ldots \beta_{15}$. The coefficients are constant for certain values of factors.

For example, the coefficient β_1 is a table relationship between the number of lanes on the road, lanes' or highway's width. β_2 depends on the roadside's width, β_3 represents the relationship between the distance from the edge of the carriageway to the obstacle and width of the lane, β_4 shows the relationship between the number of trains in the flow in % and the number of passenger cars and average trucks in %, β_5 represents a ratio indicating the relationship between the longitudinal slope in %, the length of ascent and number of trains in the flow in %. Coefficient β_6 depends on the visibility distance in meters, β_7 depends on the radius of the curve in the plan, β_8 represents a coefficient reflecting a speed limit as a sign, the ratio β_9 depends on the three parameters: the number of vehicles turning to the left in %, type of intersection and the width of the carriageway of the main road, the coefficient β_{10} depends on strengthening of the roadside, β_{11} depends on the type of road surface, β_{12} changes its value in the presence of gas stations, rest areas with a complete separation from the main road and the presence of special lanes for entry or the same attributes in the presence or absence of stripping, β_{13} depends on the type of marking, β_{14} depends on the speed limit (similar to β_8) and the presence of signs of lanes, β_{15} reveals the relationship between the number of buses in a stream in % and the number of cars in the flow in %.

1.2.1 Factors Leading to the Problem Statements in Fuzzy Conditions

The following difficulties can appear while determining of the practical capacity according to the described schemes:

1. The mentioned formulas of calculating the capacities of the road (or its part) don't take into account weather conditions, such as snowfalls, sleet, etc. In fact, in most cases it is only controlling of capacities for roads conditions in given terms, and not recalculation.
2. Repair works, traffic jams on the roads are not taken into account. Neglecting these figures leads to incorrect interpretation of the indicators included in the partial factors (lanes, highway and roadside's width).

3. Some of the indicators that make up the partial factors can be unknown due to the lack of statistical data, the unique formulations of the problems (in particular, the indicators of the number of cars in the stream, the number of vehicles turning to the left, etc.). The lack of statistical data may be caused by the construction of new roads, repairing of existed ones. It makes data collection of the number of vehicles impossible.
4. Some of the indicators included in the combined coefficients, can be misinterpreted because of the variable structure of the road system. For example, if you have statistical data about the width of the roadway, roadside, number of lanes, type of cover, the strengthening of the roadside, speed limits and carrying out repairing activities in the areas of roads, accidents, weather disasters, the information cannot be used because it is out of date and incorrect.

Thus, in spite of the existing methods of evaluating of the roads capacities, we cannot set this parameter in a crisp way due to the specificity of transportation networks, in particular the influence of weather conditions, the intervention of human activity, the errors in measurements, or lack of data about road's conditions. Consequently, the capacities of the transportation networks in below described algorithms should be represented in a fuzzy way.

Therefore, it is necessary to consider flow tasks in fuzzy conditions and develop algorithms of their solving. The main tool of solving such a tasks is fuzzy set theory [5–7].

1.3 Fuzzy Logic as the Main Tool of Dealing with Uncertainty

There are two ways of describing uncertainty inherent to the objects of the described system: randomness and fuzziness [8]. According to [9] the use of probability theory to carry out operations with fuzzy variables leads to the fact that fuzziness of any nature replaced by randomness.

Fuzziness that is the major source of uncertainty is such a type of uncertainty associated with fuzzy sets which has no clear transition from membership to non-membership of the object to the fuzzy sets. The randomness operates uncertainty using only the concept of membership or non-membership of the object to fuzzy sets.

Probability distribution of the network parameters, such as flows, arc capacities, and transmission costs is not defined due to the lack of information for the strict application of probability models and the difficulties of operating the random variables, constantly changing and expanding transport system, as well as construction of new roads. Statistical data of these parameters can be unavailable, and data acquisition can be difficult. Therefore, we cannot use the theory of probability for the most common flow tasks. Besides, there is a certain fuzziness of the estimates of the problem, the input data and vagueness of judgments in various economic, social and other systems in which a key role is given to the person (the expert).

Indeed, human nature is to think qualitatively, and fuzzy logic allows formalizing vague concepts such as "about 10," "likely, 15", "at least 100". Such network parameters as arc capacities, lower flow bounds, which are the formalization of the transportation profitability, transmission costs can not be accurately identified, measured or calculated solving flow tasks. Different factors related to the environment, errors in the measurements influence network parameters. The usage of the probability theory is incorrect due to the unique problem formulations caused by the ever-changing structure of transportation networks. Therefore, the best method to describe such problems is fuzzy logic. It is important to emphasize that the fuzzy sets theory is not designed to compete with the probability theory and statistical methods, it fills a gap in a structured uncertainty where statistics and probability cannot be applied correctly. The founder of fuzzy logic Lotfi Zadeh believed that "an excessive desire for accuracy began to have effect, nullifies the control theory and systems theory, as it leads to the fact that research in this area focus on those issues that amenable to exact solution. As a result, many classes of important problems in which data, goals and constraints are too complex and ill-defined in order to allow a precise mathematical analysis, were left aside for the reason that they are not amenable to mathematical treatment. In order to say anything substantial to the problems of this kind, we must abandon our demands of accuracy and allow results that are can be vague or uncertain" [10]. Methods based on the approach of L. Zadeh, cannot give a definitive selection criteria, their task is to drop uncompetitive, to identify the most promising ones.

Let us consider the basic concepts of the fuzzy logic.

Let us consider universal set $X = \{x\}$. *Fuzzy set* on the set X is the set of the pairs $\tilde{A} = \{ <\mu_{\tilde{A}}(x), x > | x \in X \}$, where $\mu_{\tilde{A}} : X \rightarrow [0, 1]$ is mapping of the set X into the unit interval $[0, 1]$, called membership function of the fuzzy set $\tilde{A}$. The value of the membership function $\mu_{\tilde{A}}(x)$ for the element $x \in X$ is called the membership degree. Interpretation of the membership degree is a subjective measure of the element corresponding to the concept, which meaning is formalized by the fuzzy set [11].

Fuzzy number $\tilde{B}$ is called fuzzy subset $\tilde{B}$ of the set of all real numbers X, which membership function satisfies the following conditions:

1. Continuity.
2. Normality: $\sup_{x \in X} \mu_{\tilde{B}}(x) = 1$ [12].
3. Convexity: $\mu_{\tilde{B}}(\lambda x_1 + (1 - \lambda)x_2) \geq \min(\mu_{\tilde{B}}(x_1), \mu_{\tilde{B}}(x_2))$, $x_1, x_2 \in X$, $\lambda \in [0, 1]$ [12]

One frequently operates the simplest forms of fuzzy numbers: fuzzy intervals, triangular and trapezoidal fuzzy numbers while solving practical tasks. The use of such numbers is due to the their simplicity, as well as their graphic presentation.

Fig. 1.1 Fuzzy interval

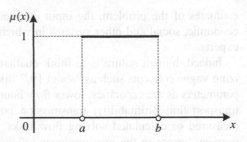

Fuzzy interval $\widetilde{A}$ is a deuce of real numbers $[a, b]$ with its membership function $\mu_{\widetilde{A}}(x)$ defined as:

$$\mu_{\widetilde{A}}(x) = \begin{cases} 1, & a \leq x \leq b \\ 0, & x < a, x < b. \end{cases} \qquad (1.5)$$

In (1.5) a—the left interval border, b—the right interval border. Graphically fuzzy interval can be represented as follows (Fig. 1.1).

Consider the basic operations carried out over the fuzzy intervals.

Let two fuzzy intervals are given $\widetilde{A}_1 = [a_1, b_1]$ и $\widetilde{A}_2 = [a_2, b_2]$.

Sum of two fuzzy intervals is [13]:

$$\widetilde{A}_1 + \widetilde{A}_2 = [a_1 + a_2, b_1 + b_2]. \qquad (1.6)$$

Comparing fuzzy intervals is performed according to [14] by comparing their deviation borders, i.e. $\widetilde{A}_1 < \widetilde{A}_2$, if $a_1 < a_2$, $b_1 < b_2$.

If this condition is not satisfied, compare fuzzy intervals by the triangular center of gravity index: $\widetilde{A}_1 > \widetilde{A}_2$, if $m(\widetilde{A}_1) < m(\widetilde{A}_2)$, where $m(\widetilde{A}_1) = (a_1 + b_1)/2$, $m(\widetilde{A}_2) = (a_2 + b_2)/2$.

If $m(\widetilde{A}_1) = m(\widetilde{A}_2)$, then the third condition is verified: if $w(\widetilde{A}_1) < w(\widetilde{A}_2)$, where $w(\widetilde{A}_1) = b_1 - a_1$ and $w(\widetilde{A}_2) = b_2 - a_2$.

The difference of two fuzzy intervals $(\widetilde{A}_1 \geq \widetilde{A}_2)$ is calculated as:

$$\widetilde{A}_1 - \widetilde{A}_2 = [a_1 - b_2, b_1 - a_2]. \qquad (1.7)$$

Operation of nonstandard subtraction of intervals is as follows [14]:

$$\widetilde{A}_1 - \widetilde{A}_2 = [\min\{a_1 - a_2, b_1 - b_2\}, \max\{a_1 - a_2, b_1 - b_2\}]. \qquad (1.8)$$

Triangular fuzzy number $\widetilde{A}$ is a triple of real numbers (a, γ, δ), with membership function $\mu_{\widetilde{A}}(x)$ defined as follows:

Fig. 1.2 Fuzzy triangular number

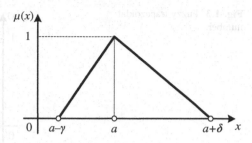

$$\mu_{\tilde{A}}(x) = \begin{cases} 1 - \frac{a-x}{\gamma}, & a - \gamma \le x \le a, \\ 1 - \frac{x-a}{\delta}, & a \le x \le a + \delta, \\ 0, & x < a - \gamma, x > a + \delta. \end{cases} \tag{1.9}$$

In (1.9) a—the center of triangular number, γ—the left deviation, δ—the right deviation. Graphically, fuzzy triangular number can be represented as follows (Fig. 1.2).

There are various ways of designations of triangular fuzzy numbers [9, 13]: (a, γ, δ) or $(a - \gamma, a, a + \delta)$. In the first case, the center and deviations are set; the second way assumes assignment of the left deviation border, the center and the right deviation border. We will use the first designation.

Consider the basic operations on triangular fuzzy numbers.

Let two triangular fuzzy numbers are set: $\tilde{A}_1 = (a_1, \gamma_1, \delta_1)$ и $\tilde{A}_2 = (a_2, \gamma_2, \delta_2)$. The sum of two triangular fuzzy numbers [15] can be represented as:

$$\tilde{A}_1 + \tilde{A}_2 = (a_1 + a_2, \gamma_1 + \gamma_2, \delta_1 + \delta_2). \tag{1.10}$$

Comparison of fuzzy triangular numbers is carried out by comparison of their centers and deviation borders, i.e. $\tilde{A}_1 < \tilde{A}_2$ if $a_1 < a_2, a_1 - \gamma_1 < a_2 - \gamma_2$, $a_1 + \delta_1 < a_2 + \delta_2$. If this condition is not satisfied, compare fuzzy triangular numbers through the triangular center of gravity: $\tilde{A}_1 < \tilde{A}_2$, if $m(\tilde{A}_1) < m(\tilde{A}_2)$, where $m(\tilde{A}_1) = (a_1 + (a_1 - \gamma_1) + (a_1 + \delta_1))/3, m(\tilde{A}_2) = (a_2 + (a_2 - \gamma_2) + (a_2 + \delta_2))/3$.

The difference of two triangular fuzzy numbers [15] $\tilde{A}_1 \ge \tilde{A}_2$ is calculated as:

$$\tilde{A}_1 - \tilde{A}_2 = (a_1 - a_2, \gamma_1 + \delta_2, \delta_1 + \gamma_2). \tag{1.11}$$

Trapezoidal fuzzy number [15] $\tilde{A}$ is a four of real numbers (a, b, γ, δ) with membership function $\mu_{\tilde{A}}(x)$ is defined as follows:

$$\mu_{\tilde{A}}(x) = \begin{cases} 1 - \frac{a-x}{\gamma}, & a - \gamma \le x < a, \\ 1, & a \le x \le b, \\ 1 - \frac{x-b}{\delta}, & b < x \le b + \delta, \\ 0. \end{cases} \tag{1.12}$$

Fig. 1.3 Fuzzy trapezoidal number

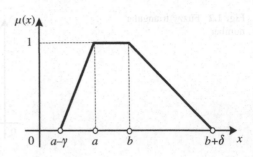

In (1.12) $[a, b]$ is called as stability interval. Graphically, some fuzzy triangular numbers can be represented as follows (Fig. 1.3).

Let two triangular fuzzy numbers are set $\widetilde{A}_1 = (a_1, b_1, \gamma_1, \delta_1)$ и $\widetilde{A}_2 = (a_2, b_2, \gamma_2, \delta_2)$.

Consider the main operations with the trapezoidal fuzzy numbers.

The sum of two trapezoidal fuzzy numbers is defined as:

$$\widetilde{A}_1 + \widetilde{A}_2 = (a_1 + a_2, b_1 + b_2, \gamma_1 + \gamma_2, \delta_1 + \delta_2). \tag{1.13}$$

Comparison of fuzzy trapezoidal numbers is carried out by comparing their confidence intervals and the deviation borders, i.e. $\widetilde{A}_1 < \widetilde{A}_2$, if $a_1 < a_2, b_1 < b_2, a_1 - \gamma_1 < a_2 - \gamma_2, b_1 + \delta_1 > b_2 + \delta_2$. If this condition is not satisfied, compare fuzzy triangular numbers through the center of gravity of trapeze: $\widetilde{A}_1 < \widetilde{A}_2$, if $m(\widetilde{A}_1) < m(\widetilde{A}_2)$, where $m(\widetilde{A}_1) = (a_1 + b_1 + (a_1 - \gamma_1) + (b_1 + \delta_1))/4$, $m(\widetilde{A}_2) = (a_2 + b_2 + (a_2 - \gamma_2) + (b_2 + \delta_2))/4$.

The difference of two trapezoidal fuzzy numbers $(\widetilde{A}_1 \geq \widetilde{A}_2)$ is calculated as follows:

$$\widetilde{A}_1 - \widetilde{A}_2 = (a_1 - b_2, b_1 - a_2, \gamma_1 + \delta_2, \delta_1 + \gamma_2). \tag{1.14}$$

α-cut (The set of α-level) of the fuzzy set $\widetilde{A}$ is crisp subset of the universal set X, which elements have membership grades, greater than or equal to $\alpha : \widetilde{A}_\alpha = \{x : \mu_{\widetilde{A}}(x) \geq \alpha\}$, $\alpha \in [0, 1]$. The value of α is called α-level.

Fuzzy intervals, which operations are given in (1.6)–(1.8), fuzzy triangular numbers with operations given in (1.10)–(1.11) and trapezoidal numbers, which operations are presented in (1.13)–(1.14) are proposed to use in the present book as a main tool of uncertainty, because it is a universal representation of fuzzy numbers. Researchers while operating fuzzy numbers of arbitrary shape, approximate their membership functions and lead to a linear representation because of the correctness and easiness of dealing with the *α-cuts* of linear functions.

Thus, using of fuzzy logic tool is necessary in flow problems arising in transportation networks, as a decision-maker can adopt solutions more sensitive to changes in environmental, to build a more or less risk-averse behavior strategies depending on the situation.

1.4 Flow Tasks in Transportation Networks

One of the first fundamental works devoted to the consideration of transportation networks is "Flows in networks" [16]. A detailed presentation of flow problems arising in networks is given in [17–19].

L. Ford and D. Fulkerson explain various optimization problems in transportation networks in the book, in particular, the problem of finding the maximum amount of flow in a network, first formulated by J. Danzig in 1951. The authors formulated and proved a theorem on the maximal flow and minimal cut asserting that the maximum flow from the source to the sink in the network is equal to the capacity of the minimal cut, proposed "labeling algorithm" to solve this problem. Let us consider this following problem. One of the first fundamental works devoted to the consideration of transport networks are "Flows in networks" [8]. A detailed presentation of streaming problems arising in transport networks, found in [12, 20, 21].

1.4.1 Maximum Flow Finding in a Network

The formal problem statement of the maximum flow finding in a network [2] can be represented as:

$$v = \sum_{x_j \in \Gamma(s)} \xi_{sj} = \sum_{x_k \in \Gamma^{-1}(t)} \xi_{kt} \to \max, \qquad (1.15)$$

$$\sum_{x_j \in \Gamma(x_i)} \xi_{ij} = \sum_{x_k \in \Gamma^{-1}(x_i)} \xi_{ki} = \begin{cases} v, & x_i = s, \\ -v, & x_i = t, \\ 0, & x_i \neq s, t, \end{cases} \qquad (1.16)$$

$$\xi_{ij} \leq u_{ij}, \quad \forall (x_i, x_j) \in A. \qquad (1.17)$$

In the model (1.15)–(1.17) ξ_{ij}—the flow value passing along the arc (x_i, x_j); v—the maximum flow value in the network; s—the initial node of the graph (the source); t—the terminal node (the sink); $\Gamma(x_i)$—the set of nodes, arcs from the node $x_i \in X$ go to; $\Gamma^{-1}(x_i)$—the set of nodes, arcs from the node $x_i \in X$ go from; u_{ij}—the maximum amount of flow that can pass through the arc (x_i, x_j) (arc capacity). Equation (1.15)—objective function means that we have to maximize the total number of units v going from the source, which is equal to the total number of flow units entering the sink. Equation (1.16) is the condition of the flow conservation, which indicates that the amount of flow entering any node x_i besides the source and the sink equals to the amount of flow leaving any node x_i besides the source and the sink. Arc flow constraints are given in Eq. (1.17).

The urgency of this problem is in the fact that it allows finding the maximum volume of the traffic between two given stations on the road map. Finding the maximum flow value under the restrictions on the arc capacities allows you to find the road sections with saturated traffic, redistribute it correctly and move it from a busy area to a free one.

This problem has also been considered by the authors [22–25]. In the works of the authors Christofides [2], Minieka [17], Hu [26] solution for the problem of the maximum flow finding in the transportation network based on the "labeling method" is offered.

There are various modifications of the Ford and Fulkerson's "labeling method" [19]. Among them there are algorithm proposed in 1972 by Edmonds and Karp [27], where the shortest augmenting path from the source to the sink in the residual network is chosen at each step. The shortest route can be found by the "breadth first search". Other scholars, such as Diniz [28], Karzanov [29], Cherkasky [30, 31] worked to improve the running time of the algorithm and reduction of complexity either. The most modern modification of the Ford-Fulkerson's algorithm is Orlin's algotithm [32].

In practice, the most commonly used algorithm is an algorithm of J. Edmonds and M. Karp with running time $O(XA^2)$, where где X—the set of nodes, and A—the set of arcs.

Unlike L. Ford and D. Fulkerson's algorithm, where the flow transportation occurs along any augmenting path and searching a path is carried out by the labeling method, considered algorithm passes the flow along the shortest path in terms of the number of arcs More recent modifications of the algorithm are used less because of the lack of description in the flow literature.

Considering the fuzzy nature of the network parameters, in particular, the arc capacities it is required to consider algorithms of the maximum flow finding in the transportation network in fuzzy terms [33]. The maximum flow finding in networks in fuzzy conditions was weakly reported in the literature. In [34] a solution of this problem based on interval arc capacities was proposed. The same authors have proposed to solve this problem using so-called "fuzzy graphs" [35]. There are modern articles on the same subject, which represent the solution of this problem by the simplex method of linear programming [36, 37]. But this method does not fully operate the truth transportation network. Graph methods of solving the problems of the maximum flow finding in fuzzy conditions considered in the flow literature search the maximum flow by L. Ford and D. Fulkerson's algorithm, i.e. selecting any ways of increasing the flow without considering the faster algorithms of searching the ways of passing the flows. Thus, it is necessary to consider the algorithm for the maximum flow finding in the transportation network in fuzzy conditions [20] using a more efficient method for searching the augmenting path.

1.4.2 Maximum Flow Finding in a Network with Nonzero Lower Flow Bounds

The task of the maximum flow finding is variation of the more complex problem of the maximum flow determining in the network with nonzero lower flow bounds . Lower flow bound is the minimal feasible flow that must be passed along the corresponding arc. Let's assume a network consists of railways, sea and air roads. Therefore, the freight trains have certain level of load, which is more than profitability threshold, transport planes don't fly at a low load. Thus, it is necessary to introduce nonzero lower flow bounds, which can lead to the absence of feasible flow. The introduction of the lower flow bounds makes existing the feasible flow non-obvious.

The formal problem statement of the maximum flow finding task taking into account lower bounds can be represented as follows:

$$v = \sum_{x_j \in \Gamma(s)} \xi_{sj} = \sum_{x_k \in \Gamma^{-1}(t)} \xi_{kt} \rightarrow \max, \tag{1.18}$$

$$\sum_{x_j \in \Gamma(x_i)} \xi_{ij} = \sum_{x_k \in \Gamma^{-1}(x_i)} \xi_{ki} = \begin{cases} v, & x_i = s, \\ -v, & x_i = t, \\ 0, & x_i \neq s, t, \end{cases} \tag{1.19}$$

$$l_{ij} \leq \xi_{ij} \leq u_{ij}, \quad \forall (x_i, x_j) \in A. \tag{1.20}$$

In (1.18)–(1.20) l_{ij}—the lower flow bound of the arc (x_i, x_j) of the network.

The task of the maximum flow finding with nonzero lower flow bounds was not widely reflected in the literature. In particular, the authors Chritsofides [2], Mutry and others [21, 38] consider the problem of maximum flow determining in the transportation network, taking into account the lower and upper flow bounds. Thus, Christofides [2] considers the problem of the feasible flow finding, i.e. does not actually solves the maximum flow problem with lower and upper flow bounds in the graph. This problem was not considered in the literature in fuzzy conditions. Therefore, it is necessary to develop algorithm of the maximum flow finding with nonzero lower flow bounds in the network taking into account the uncertainty of the network.

1.4.3 Minimum Cost Flow Finding in a Network

The major problem arising in the study of flows is the problem of determining the minimum cost flow. This task is observed in the works of Ford and Fulkerson [16], Minieka [17], Christofides [2], Phillips and Garcia-Diaz [39], Busacker and Gowen [40] and Klein [41]. The essence of this task is to determine the minimum cost transportation of a certain number of the flow units in a transportation network. Formally, the present problem statement can be formulated as:

$$\sum_{(x_i,x_j)\in A} c_{ij}\xi_{ij}, \tag{1.21}$$

$$\sum_{x_j\in \Gamma(x_i)} \xi_{ij} = \sum_{x_k\in \Gamma^{-1}(x_i)} \xi_{ki} = \begin{cases} \rho, & x_i = s, \\ -\rho, & x_i = t, \\ 0, & x_i \neq s, t, \end{cases} \tag{1.22}$$

$$\xi_{ij} \leq u_{ij}, \quad \forall\, (x_i, x_j) \in A. \tag{1.23}$$

In (1.21)–(1.23) c_{ij}—transmission cost of the one flow unit along the arc (x_i, x_j); ξ_{ij}—the flow value, passing along the arc (x_i, x_j); ρ—the given flow value not more than the maximum flow in the network; s—the initial node of the network (the source); t—the terminal node of the network (the sink); u_{ij}—capacity of the arc (x_i, x_j) of the network. Equation (1.21)—the objective function reflecting the minimum cost flow finding, Eq. (1.22) is condition of the flow conservation. Arc flow constraints are given in Eq. (1.23).

Variation of this problem is the minimum cost maximum flow finding task, wherein it is necessary to determine the minimum cost way of sending the maximum flow. The model of this problem is similar to the model (1.21)–(1.23), taking into account the observation that v is the maximum flow in a graph:

$$\sum_{(x_i,x_j)\in A} c_{ij}\xi_{ij}, \tag{1.24}$$

$$\sum_{x_j\in \Gamma(x_i)} \xi_{ij} = \sum_{x_k\in \Gamma^{-1}(x_i)} \xi_{ki} = \begin{cases} v, & x_i = s, \\ -v, & x_i = t, \\ 0, & x_i \neq s, t, \end{cases} \tag{1.25}$$

$$\xi_{ij} \leq u_{ij}, \quad \forall\, (x_i, x_j) \in A. \tag{1.26}$$

The methods of the minimum cost flow finding (1.24)–(1.26) [42] in networks in fuzzy conditions can be divided into graph techniques [42] and the methods of linear programming [43]. In particular, solutions by the graph methods are considered in [2, 16, 17, 39]. The advantages of this approach are great visualization and less cumbersome. The minimum cost flow is proposed to find by Busacker– Gowen and Klein's algorithms in [26]. The first algorithm is in the finding of the shortest paths (the minimum cost paths) and pushing the flows along them from the zero flow until the chain ceases to be the shortest one. Other words, we send the maximum flow value from the source to the sink with zero total transmission cost of each unit, then—send the maximum flow value from the source to the sink with total transmission cost of each unit equals to 1. Each time perform a similar procedure increasing the total transmission cost of the one flow unit by the one unit. The algorithm terminates in two cases: (1) when the given flow value is transferred from the source to the sink, or (2) when any additional unit of flow cannot be passed. M. Klein's algorithm starts with a random flow distribution, consequently checking whether distribution is optimal. The optimality criterion is the absence of negative cost cycles in the considered graph. Floyd's algorithm [44, 45] is proposed to use as subalgorithm for searching of negative cost cycles. If the graph has no negative cost cycle, the current flow distribution has a minimum cost; if a negative cost cycle is found, push $\delta = \min[u_{ij}]$ units along it, i.e. the minimum of arc capacities of the cycle. Then build a new graph with the new flow values until we get an optimal distribution (i.e. the absence of the negative cost cycles in the current graph).

In [2, 17, 39] the investigated problem is considered as LP problem. This approach is less visual, since researcher solves the problem indirectly, without performing graph operations, but it allows to solve the problem in a universal way.

In real life, to solve the flow problems excluding changes in the environment and human activities is impossible, because network parameters such as arc capacities and transmission costs cannot be accurately measured according to their nature. Then we turn to the minimum cost flow problem in a transportation network in fuzzy conditions.

Methods for solving the minimum cost flow problem in a network in fuzzy conditions can be divided into two classes. The first class presents the use of fuzzy LP that was widely reported in the literature [46, 47]. The authors [36, 48–50] consider the problem of "fully fuzzy LP". These tasks are cumbersome and cannot provide an optimal solution when determining the minimum flow value.

In [51] solving of fuzzy LP tasks dealing with ranking functions [52, 53] is considered. In order to consider alternative solutions that may arise in similar problems, algorithm described in [11] is applied. Solution of the minimum cost flow problem based on the ranking methods with fuzzy values of costs and crisp arc capacities is presented in [54]. There are approaches that convert the problem into one-criterion from two or three-criterion etc. depending on whether the coefficients are defined as intervals, fuzzy triangular numbers etc. These approaches are applied when the fuzzy coefficients appear in the objective function. One way for solving such tasks is obtaining Pareto-optimal solution of n-criteria problem, depending on the parameter λ [55]. These approaches are described only for intervals. The disadvantage of these methods is time-consuming. The second class involves the use of conventional flow algorithms for determining the minimum cost flow, which operate fuzzy data instead of crisp values and "blur" them according the certain rules in the end of the algorithm [56, 57]. But in papers devoted to finding a minimum cost flow in fuzzy conditions no attention is paid to the method of the shortest path choice and there is no distinction between the minimum cost flow algorithms and maximum flow minimum cost algorithms. Therefore, it is necessary to develop algorithm for minimum cost flow finding in terms of fuzzy arc capacities and transmission costs and to justify the choice of the fuzzy shortest path algorithm.

1.4.4 Minimum Cost Flow Finding in a Network with Nonzero Lower Flow Bounds

The problem of the minimum cost flow finding is a special case of the more general problem of the minimum cost flow finding with nonzero lower flow bounds. The formal statement of the problem can be represented as:

$$\sum_{(x_i, x_j) \in A} c_{ij} \xi_{ij}, \tag{1.27}$$

$$\sum_{x_j \in \Gamma(x_i)} \xi_{ij} = \sum_{x_k \in \Gamma^{-1}(x_i)} \xi_{ki} = \begin{cases} \rho, & x_i = s, \\ -\rho, & x_i = t, \\ 0, & x_i \neq s, t, \end{cases} \tag{1.28}$$

$$l_{ij} \leq \xi_{ij} \leq u_{ij}, \quad \forall (x_i, x_j) \in A. \tag{1.29}$$

The urgency of the problem (1.27)–(1.29) is that the cheapest routes of the flow transportation are found taking into account economic feasibility of carriage by assigning nonzero lower flow bounds. Just the problem of the minimum cost circulation finding is considered in the literature [58, 59], but this problem involves the condition of the flow conservation for all nodes, i.e. there are no selected nodes in graphs for such problems. The problem of the minimum cost flow finding with nonzero lower flow bounds was not discussed in fuzzy conditions, therefore, it is necessary to develop an algorithm of this problem in a fuzzy transportation network.

A variation of this problem is the minimum cost maximum flow problem with nonzero lower flow bounds, wherein it is necessary to determine the minimum cost way of sending the maximum flow with arc capacities constraints. The model of this problem is similar to the model (1.27)–(1.29), taking into account that v is the maximum flow in a graph:

$$\sum_{(x_i, x_j) \in A} c_{ij} \xi_{ij}, \tag{1.30}$$

$$\sum_{x_j \in \Gamma(x_i)} \xi_{ij} = \sum_{x_k \in \Gamma^{-1}(x_i)} \xi_{ki} = \begin{cases} v, & x_i = s, \\ -v, & x_i = t, \\ 0, & x_i \neq s, t, \end{cases} \tag{1.31}$$

$$l_{ij} \leq \xi_{ij} \leq u_{ij}, \quad \forall (x_i, x_j) \in A. \tag{1.32}$$

The present task (1.30)–(1.32) was not considered in fuzzy conditions, therefore, it is necessary to develop the formal algorithm for its solving.

1.5 Flow Tasks in Dynamic Networks

Conventional flow tasks observed earlier assume the instant flow, passing along the arcs of the graph, what certainly, is simplification of the real life. Such tasks are called static flow tasks. In fact, it turns out that the flow spends certain time passing along the arcs of the graph. Then, we turn to dynamic networks, in which each flow unit passes from the source to the sink for a period of time less than given. Dynamic networks describe complex systems, problems of decision-making, models, which parameters can vary over time. Such models can be found in communication systems, economic planning, transportation systems and many other applications, so they have a wide practical application. Flow tasks in dynamic networks allow to find the maximum value of the transportation and choose the routes of the minimum cost for transportation a certain flow units with the limited flow range. Consider the notion of the dynamic network according to the Definition 1.4.

Definition 1.4 Dynamic network is a network [16] $G = (X, A)$, where $X = \{x_1, x_2, \ldots, x_n\}$—the set of nodes, $A = \{(x_i, x_j)\}$, $i, j \in I = \overline{1, n}$—the set of arcs. Each arc of the dynamic graph (x_i, x_j) is set two parameters: transit time τ_{ij} and arc capacity u_{ij}. The time horizon $T = \{0, 1, \ldots, p\}$ determining that all flow units sent from the source must arrive at the sink within the time p is given [16]. Let τ_{ij} be a positive number. Let $\Gamma(x_i)$ is the set of nodes, arcs from x_i go to, $\Gamma^{-1}(x_i)$ is the set of nodes, arcs from x_i go from. Thus, not more than u_{ij} units of flow can be sent along the arc (x_i, x_j) at each time period in dynamic networks. Let x_j be the terminal node and x_i is the initial node of the arc (x_i, x_j), then the flow leaving x_i at $\theta \in T$ will enter x_j at time period $\theta + \tau_{ij}$.

There are various flow problem statements in dynamic graphs: the maximum dynamic flow finding, the minimum cost dynamic flow finding, etc. Let us consider these problem statements in details.

1.5.1 Maximum Flow Finding in Dynamic Networks with Zero and Nonzero Lower Flow Bounds

Historically, the maximum flow finding in dynamic graph was the first task flow task in dynamic graphs, described in the literature. The notion "dynamic flow" was proposed by Ford and Fulkerson [16] as a task of maximum dynamic flow finding in a network. This problem is in finding of maximum flow, passing from the source (s) to the sink (t), $s, t \in X$ in the network for p discrete time periods, starting from zero period of time. All flow units leaving the source must arrive at the sink not later than at p. Let $v(p)$ be the total number of flow units leaving the source and entering the sink for time periods 0,..., p. This task can be formulated as follows [16, 60, 61]:

$$\textit{Maximize } v(p) \tag{1.33}$$

$$\sum_{\theta=0}^{p} \left(\sum_{x_j \in \Gamma(x_i)} \xi_{ij}(\theta) - \sum_{x_j \in \Gamma^{-1}(x_i)} \xi_{ji}(\theta - \tau_{ji}) \right) = v(p), \; x_i = s, \tag{1.34}$$

$$\sum_{\theta=0}^{p} \left(\sum_{x_j \in \Gamma(x_i)} \xi_{ij}(\theta) - \sum_{x_j \in \Gamma^{-1}(x_i)} \xi_{ji}(\theta - \tau_{ji}) \right) = 0, \; x_i \neq s, t; \; \theta, \theta - \tau_{ji} \in T, \tag{1.35}$$

$$\sum_{\theta=0}^{p} \left(\sum_{x_j \in \Gamma(x_i)} \xi_{ij}(\theta) - \sum_{x_j \in \Gamma^{-1}(x_i)} \xi_{ji}(\theta - \tau_{ji}) \right) = -v(p), \; x_i = t, \tag{1.36}$$

$$0 \leq \xi_{ij}(\theta) \leq u_{ij}, \quad \forall \, (x_i, x_j) \in A, \; \theta \in T. \tag{1.37}$$

In the model (1.33)–(1.37) the expression (1.33) means that we maximize the amount of flow v for p periods of time. Expression (1.34) shows that the maximum flow v for p periods is equal to the total flow $\sum_{\theta=0}^{p} \sum_{x_j \in \Gamma(x_i)} \xi_{ij}(\theta)$, $x_i = s$ leaving the source for p periods of time. Expression (1.36) reflects that maximum flow v for p periods is equal to the total flow $\sum_{\theta=0}^{p} \sum_{x_j \in \Gamma^{-1}(x_i)} \xi_{ji}(\theta - \tau_{ji})$, $x_i = t$ entering the sink for p periods. The total flow $\sum_{\theta=0}^{p} \sum_{x_j \in \Gamma^{-1}(x_i)} \xi_{ji}(\theta - \tau_{ji})$, $x_i = s$ entering the source for p periods is equal to the total flow leaving the sink $\sum_{\theta=0}^{p} \sum_{x_j \in \Gamma(x_i)} \xi_{ij}(\theta)$, $x_i = t$ for p periods and is equal to zero. The total amount of flow units $\sum_{\theta=0}^{p} \sum_{x_j \in \Gamma^{-1}(x_i)} \xi_{ji}(\theta - \tau_{ji})$, $x_i \neq s, t$ entering the node x_i at time periods $(\theta - \tau_{ji})$ for p periods equals to the amount of flow units $\sum_{\theta=0}^{p} \sum_{x_j \in \Gamma(x_i)} \xi_{ij}(\theta)$, $x_i \neq s, t$ leaving the node x_i at time periods θ for p periods for each period of time θ and for each node x_i, except the source and the sink, as stated in (1.35). Inequality (1.37) indicates that the flows for all time periods $\xi_{ij}(\theta)$ should be less than arc capacities of the corresponding arcs.

The task of maximum dynamic flow finding was widely reported in the literature. L. Ford and D. Fulkerson proposed two methods for its solution: the first is based on the constructing of the "time-expanded network" [16, 62], and the second— implements the algorithm of the minimum cost flow finding considering the flow transit time along the arc equals to the arc cost of the corresponding arc, applying the shortest path algorithm.

Minieka, Aronson [17, 60] studied "stationary-dynamic" models of the maximum flows in addition to Ford and Fulkerson [16]. In particular, Minieka [17] simulated a situation, in which some of the arcs of the dynamic graph may not be available at some time periods. In [63–65] the solution of the maximum flow task in a dynamic network in the case of either discrete or continuous time periods are considered.

The problem statement of the maximum dynamic flow finding with nonzero lower flow bounds is equivalent to the problem statement (1.33)–(1.37) with the condition of replacing restrictions on the arc capacities (1.37) on the following expression:

$$l_{ij} \leq \xi_{ij}(\theta) \leq u_{ij}, \quad \forall (x_i, x_j) \in A, \ \theta \in T. \tag{1.38}$$

The present problem allows finding the maximum value of cargo traffic between the selected points in the road map with given nonzero transmission profitability lower bounds, as shown in (1.38) for given time period. Fonoberova, Lozovanu [66] investigated the problems of the maximum dynamic flow finding, in particular, the problem of admissibility of flow existence in the case, when arcs of the dynamic network have nonzero lower flow bounds. The computational complexity of this algorithm depending on the complexity of the chosen method of the maximum flow determining in the time-expanded version of the original graph was shown by the same authors. Researches in the field of dynamic flows allow taking into account either discrete or continuous time intervals [67].

In its turn the problems of the maximum flow finding with nonzero and zero lower flow bounds were not considered in fuzzy conditions, although such parameters of the transportation network, as the lower and upper flow bounds cannot be accurately measured. These tasks require either the equality of the flow bounds for all time periods, without actually considering the possibility of these parameters change over time. Described remarks make it necessary to consider tasks of the maximum flow finding with zero and nonzero lower flow bounds in fuzzy conditions and in terms of network parameters depending on the time.

1.5.2 Minimum Cost Flow Finding in Dynamic Network with Zero and Nonzero Lower Flow Bounds

The more difficult area of research is the minimum cost dynamic flow finding. The urgency of these problems is that their solution allows finding the optimal routes of the cargo transportation for given number of time periods. Problem statement of the minimum cost flow finding in a dynamic graph $G = (X, A)$ [68] can be represented as follows:

$$\sum_{\theta=0}^{p} \sum_{(x_i, x_j) \in A} c_{ij} \xi_{ij}(\theta), \quad \forall (x_i, x_j) \in A, \ \theta \in T, \tag{1.39}$$

$$\sum_{\theta=0}^{p} \left(\sum_{x_j \in \Gamma(x_i)} \xi_{ij}(\theta) - \sum_{x_j \in \Gamma^{-1}(x_i)} \xi_{ji}(\theta - \tau_{ji}) \right) = \rho(p), \ x_i = s, \tag{1.40}$$

$$\sum_{\theta=0}^{p}\left(\sum_{x_j\in\Gamma(x_i)}\xi_{ij}(\theta)-\sum_{x_j\in\Gamma^{-1}(x_i)}\xi_{ji}(\theta-\tau_{ji})\right)=0,\ x_i\neq s,t;\ \theta,\theta-\tau_{ji}\in T,\quad(1.41)$$

$$\sum_{\theta=0}^{p}\left(\sum_{x_j\in\Gamma(x_i)}\xi_{ij}(\theta)-\sum_{x_j\in\Gamma^{-1}(x_i)}\xi_{ji}(\theta-\tau_{ji})\right)=-\rho(p),\ x_i=t,\quad(1.42)$$

$$0\leq\xi_{ij}(\theta)\leq u_{ij},\ \forall\,(x_i,x_j)\in A,\ \theta\in T.\quad(1.43)$$

In the model (1.39)–(1.43) c_{ij}—the transmission cost of the one flow unit along the arc (x_i, x_j) of the network, $\xi_{ij}(\theta)$—the flow leaving the node x_i at the time period θ, $\rho(p)$—the given flow value for p time periods.

The variation of this problem is the task of the minimum cost maximum flow finding in a dynamic network [68], represented as:

$$\sum_{\theta=0}^{p}\sum_{(x_i,x_j)\in A}c_{ij}\xi_{ij}(\theta),\ \forall\,(x_i,x_j)\in A,\ \theta\in T,\quad(1.44)$$

$$\sum_{\theta=0}^{p}\left(\sum_{x_j\in\Gamma(x_i)}\xi_{ij}(\theta)-\sum_{x_j\in\Gamma^{-1}(x_i)}\xi_{ji}(\theta-\tau_{ji})\right)=v(p),\ x_i=s,\quad(1.45)$$

$$\sum_{\theta=0}^{p}\left(\sum_{x_j\in\Gamma(x_i)}\xi_{ij}(\theta)-\sum_{x_j\in\Gamma^{-1}(x_i)}\xi_{ji}(\theta-\tau_{ji})\right)=0,\ x_i\neq s,t;\ \theta,\theta-\tau_{ji}\in T,\quad(1.46)$$

$$\sum_{\theta=0}^{p}\left(\sum_{x_j\in\Gamma(x_i)}\xi_{ij}(\theta)-\sum_{x_j\in\Gamma^{-1}(x_i)}\xi_{ji}(\theta-\tau_{ji})\right)=-v(p),\ x_i=t,\quad(1.47)$$

$$0\leq\xi_{ij}(\theta)\leq u_{ij},\ \forall\,(x_i,x_j)\in A,\ \theta\in T.\quad(1.48)$$

In the model (1.44)–(1.48) $v(p)$—is the given flow value for p time periods.

This problem was considered by Fleischer and Scutella [67] proved that ability of intermediate nodes to maintain the flow doesn't lead to reducing the cost of dynamic flow. Turning to the "time-expanded" graph for the minimum cost flow finding is represented in [69]. As was mentioned, the arc capacities and flow transit times along the real network arcs can vary depending on the time, so the dynamic models are reflected in the articles of the authors [70–73], which deal with the transit network parameters Time parameters. Authors Chalmet et al. [74] used the algorithm of the maximum dynamic flow finding to evacuate buildings. As a result, the algorithm receives the minimum time for which one can evacuate a large building.

Subtask of the minimum cost flow finding in the dynamic network is the problem of the shortest paths finding in the dynamic graph. The purpose of this task is to define the dynamic path of the minimal length for each node i from the source s of the network. For the first time this task was introduced by Cooke and Halsey [75] and has been widely studied in the literature by the authors [76–81] with the assumption that all flow transit times along the arcs are non-negative.

Despite the investigations of the authors in the field of dynamic networks described problems were not actually considered in terms of fuzziness. Authors Glockner et al. [82] considered this problem as a linear programming problem. The minimum cost flow finding in the fuzzy dynamic network with nonzero lower flow bounds neither was considered in the literature. Thus, it becomes necessary to develop algorithms of the maximum flow finding with zero and nonzero lower flow bounds, the minimum cost, subtask of minimizing the maximum flow and the minimum cost flow taking into account the nonzero lower flow bounds in a fuzzy dynamic network according to its transit structure.

1.6 Summary

This chapter deals with main concepts of fuzzy logic. The basic flow tasks in fuzzy networks are described. The factors leading to the problem statements in fuzzy conditions are given. The conclusion is in necessity of flow tasks considering in fuzzy conditions.

References

1. Frank H, Frisch IT (1971) Communication, transmission, and transportation networks. Addison Wesley, Boston
2. Christofides N (1975) Graph theory: an algorithmic approach. Academic Press, New York
3. Bozhenyuk A, Gerasimenko E (2014) Flows finding in networks in fuzzy conditions. In: Kahraman C, Öztaysi B (eds) Supply chain management under fuzzines, studies in fuzziness and soft computing, vol 313, part III. Springer, Berlin, pp 269–291. doi:10.1007/978-3-642-53939-8_12
4. Rukovodstvo po otsenke propusknoy sposobnosti avtomobilnikh dorog (1982). Minavtodor RSFSR, Transport, Moskva
5. Kaufmann A (1975) Introduction to the theory of fuzzy subsets, vol 1. Academic Press, New York
6. Kofman A, Hilo Aluja H (1992) Introduction the theory of fuzzy sets in enterprise management. High School, Minsk
7. Zadeh LA, Fu KS, Tanaka K, Shimura M (eds) (1975) Fuzzy sets and their applications to cognitive and decision processes. Academic Press, New York
8. Bellman RE, Zadeh LA (1970) Decision making in a fuzzy environment. Manag Sci 17: 141–164

9. Borisov AN, AlekseevA V, Merkuryeva GV, Slyadz NN, Glushkov VI (1989) Fuzzy information processing in decision-making systems. Radio and Communication Publisher, Moscow
10. Zadeh LA (1975) The concept of a linguistic variable and its application to approximate reasoning. Inf Sci 8(part I):199–249; Inf Sci 8(part II):301–335; Inf Sci 9(part III):43–80
11. Borisov AN, Alekseev AV, Krumberg OA et al (1982) Decision making models based on linguistic variable. Zinatne, Riga
12. Zimmermann HJ (1991) Fuzzy set theory and its applications, 2nd edn. Kluwer Academic Publishers, Boston
13. Bozhenyuk A, Rozenberg I, Starostina T (2006) Analiz i issledovaniye potokov i zhivuchesti v transportnykx setyakh pri nechetkikh dannykh. Nauchnyy Mir, Moskva
14. Bershtein LS, Dzuba TA (1998) Construction of a spanning subgraph in the fuzzy bipartite graph. In: Proceedings of EUFIT'98, Aachen, pp 47–51
15. Dubois D, Prade H (1978) Operations on fuzzy numbers. Int J Syst Sci 9(6):613–626
16. Ford LR, Fulkerson DR (1962) Flows in networks. Princeton University Press, Princeton
17. Minieka E (1978) Optimization algorithms for networks and graphs. Marcel Dekker Inc, New York and Basel
18. Goldberg AV, Tardos E, Tarjan RE (1990) Network flow algorithms. In: Korte B, Lovasz L, Proemel HJ, Schrijver A (eds) Paths, flows and VLSI-design. Springer, Berlin, pp 101–164
19. Ahuja RK, Magnanti TL, Orlin JB (1993) Network flows: theory, algorithms, and applications. Prentice-Hall Inc, New Jersey
20. Bozhenyuk A, Rozenberg I, Rogushina E (2011) The method of the maximum flow determination in the transportation network in fuzzy conditions. In: Proceedings of the congress on intelligent systems and information technologies «IS&IT'11», vol 4. Physmathlit, Moscow, pp 17–24 (Scientific Publication in 4 volumes)
21. Murty KG (1992) Network programming. Prentice Hall, Upper Saddle River
22. Even S (1979) Graph algorithms. Computer Science Press, Potomac
23. Lawler EL (1976) Combinatorial optimization: networks and matroids. Holt, Rinehart, and Winston, New York
24. Papadimitriouc H, Steiglitz K (1982) Combinatorial optimization: algorithms and complexity. Prentice-Hall, Englewood Cliffs
25. Tarjan RE (1983) Data structures and network algorithms. Society for Industrial and Applied Mathematics, Philadelphia
26. Hu TC (1969) Integer programming and network flows. Addison Wesley, Boston
27. Edmonds J, Karp RM (1972) Theoretical improvements in algorithmic efficiency for network flow problems. J Assoc Comput Mach 19(2):248–264
28. Dinic EA (1970) Algorithm for solution of a problem of maximal flow in a network with power estimation. Soviet Math Doklady 11:1277–1280
29. Karzanov AV (1974) Determining the maximum flow in the network by the method of preflows. Soviet Math Doklady 15:434–437
30. Cherkassky RV (1977) Algorithms of construction of maximal flow in networks with complexity $O(n^2\sqrt{m})$ operations. Math Methods Solut Econ Probl 7:112–125
31. Cherkassky BV (1979) A fast algorithm for computing maximum flow in a network. In: Karzanov AV (eds) Collected papers, combinatorial methods for flow problems. The Institute for Systems Studies, Moscow, vol 3, pp 90–96 (English translation appears in AMS Trans 158:23–30 (1994))
32. Orlin JB (2013) Max flows in O(nm) time or better. In: Proceedings of the 2013 symposium on the theory of computing, pp 765–774
33. Bozhenyuk AV, Rozenberg IN, Rogushina EM (2011) Approach of maximum flow determining for fuzzy transportation network. Izvestiya SFedU Eng Sci 5(118):83–88
34. Chanas S, Kolodziejczyk W (1982) Maximum flow in a network with fuzzy arc capacities. Fuzzy Sets Syst 8(2):165-153

35. Chanas S, Delgado M, Verdegay JL, Vila M (1995) Fuzzy optimal flow on imprecise structures. Eur J Oper Res 83(3):568–580
36. Kumar A, Kaur J, Singh P (2010) Fuzzy optimal solution of fully fuzzy linear programming problems with inequality constraints. Int J Appl Math Comput Sci 6:37–41
37. Kumar A, Kaur J (2011) Solution of fuzzy maximal flow problems using fuzzy linear programming. Int J Comput Math Sci 5(2):62–66
38. Yi T, Murty KG (1991) Finding maximum flows in networks with nonzero lower bounds using Preflow methods. In: Technical report, IOE Department, University of Michigan, Ann Arbor, Mich
39. Garcia-Diaz A, Phillips DT (1981) Fundamentals of network analysis. Prentice-Hall, Englewood Cliffs
40. Busacker RG, Gowen P (1961) A procedure for determining a family of minimum-cost network flow patterns. In: Technical report 15, operations research office, John Hopkins University
41. Klein M (1967) A primal method for minimal cost flows with applications to the assignment and transportation problems. Manage Sci 14:205–220
42. Kovács P (2015) Minimum-cost flow algorithms: an experimental evaluation. Optim Methods Softw 30(1):94–127
43. Ahuja RK, Goldberg AV, Orlin JB, Tarjan RE (1992) Finding minimum-cost flows by double scaling. Mathematical Programming 53:243–266
44. Warshall S (1962) A theorem on boolean matrices. J ACM 9(1):11–12
45. Floyd RW (1962) Algorithm 97: shortest path. Commun ACM 5(6):345
46. Ganesan K, Veeramani P (2006) Fuzzy linear programs with trapezoidal fuzzy numbers. Ann Oper Res 143:305–315
47. Thakre PA, Shelar DS, Thakre SP (2009) Solving fuzzy linear programming problem as multi objective linear programming problem. In: Proceedings of the world congress on engineering (WCE 2009), London, U.K
48. Dutta D, Murthy AS (2010) Multi-choice goal programming approaches for a fuzzy transportation problem. IJRRAS 2(2):132–139
49. Kumar A, Singh P, Kaur J (2010) Generalized simplex algorithm to solve fuzzy linear programming problems with ranking of generalized fuzzy numbers. Turkish J Fuzzy Syst (TJFS) 1(2):80–103
50. Maleki HR, Tata M, Mashinchi M (2000) Linear programming with fuzzy variables. Fuzzy Sets Syst 109:21–33
51. Maleki HR, Mashinchi M (2004) Fuzzy number linear programming: a probabilistic approach. J Appl Math Comput 15:333–341
52. Yoon KP (1996) A probabilistic approach to rank complex fuzzy numbers. Fuzzy Sets Syst 80:167–176
53. Allahviranloo T, Lotfi FH, Kiasary MK, Kiani NA, Alizadeh L (2008) Solving fully fuzzy linear programming problem by the ranking function. Appl Math Sci 2(1):19–32
54. Gerasimenko EM (2012) Minimum cost flow finding in the network by the method of expectation ranking of fuzzy cost functions. Izvestiya SFedU Eng Sci 4(129):247–251
55. Chanas S, Kuchta D (1994) Linear programming problem with fuzzy coefficients in the objective function. In: Delgado M et al (eds) Fuzzy Optimization, Physica-Verlag, Heidelberg, pp 148–157
56. Malyshev NG, Bershteyn LS, Bozhenyuk AV (1991) Nechetkiye modeli dlya ekspertnykh sistem v SAPR. Energoatomizdat, Moscow
57. Bozhenyuk A, Gerasimenko E (2013) Methods for maximum and minimum cost flow determining in fuzzy conditions. World Appl Sci J 22:76–81. doi:10.5829/idosi.wasj.2013.22. tt.22143 (Special Issue on Techniques and Technologies)
58. Tardos E (1985) A strongly polynomial minimum cost circulation algorithm. Combinatorica 5 (3):247–255
59. Goldberg AV, Tarjan RE (1989) Finding minimum-cost circulations by canceling negative cycles. J ACM 36(4):873–886

60. Aronson JE (1989) A survey of dynamic network flows. Ann Oper Res 20:1–66
61. Ciurea E (1997) Two problems concerning dynamic flows in node and arc capacitated networks. Comput Sci J Moldova 5(3):15, 298–308
62. Silver MR, de Weck O (2006) Time-expanded decision network: a new framework for designing evolvable complex systems. In: Proceedings of 11th AIAA/ISSMO multidisciplinary analysis and optimization conference 6–8 September, Portsmouth, Virginia, pp 1–15
63. Lovetskii S, Melamed I (1987) Dynamic flows in networks. Autom Remote Control 48 (11):1417–1434 (Translated from Avtomatika i Telemekhanika (1987), vol 11, pp 7–29)
64. Powell W, Jaillet P, Odoni A (1995) Stochastic and dynamic networks and routing. In: Ball MO, Magnanti TL, Monma CL, Nemhauser GL (eds) Handbook in operations research and management science, vol 8. Network Routing, Elsevier, Amsterdam, pp 141–296
65. Orlin JB (1983) Maximum-throughput dynamic network flows. Mathematical Programming 27:214–231
66. Fonoberova MA, Lozovanu DD (2004) The maximum flow in dynamic networks. Comput Sci J Moldova 12(3):36, 387–396
67. Fleischer L, Skutella M (2003) Minimum cost flows over time without intermediate storage. In: Proceedings of the 35th ACM/SIAM symposium on discrete algorithms (SODA), pp 66–75
68. Kotnyek B (2003) An annotated overview of dynamic network flows. In: Technical report, INRIA, Paris, No 4936, http://hal.inria.fr/inria-00071643/
69. Orlin JB (1984) Minimum convex cost dynamic network flows. Math Oper Res 9(2):190–207
70. Gale D (1959) Transient flows in networks. Michigan Math J 6:59–63
71. Halpern J (1979) A generalized dynamic flows problem. Networks 9:133–167
72. Cai X, Sha D, Wong CK (2001) Time-varying minimum cost flow-problems. Eur J Oper Res 131:352–374
73. Nasrabadi E, Hashemi SM (2010) Minimum cost time-varying network flow problems. Optim Methods Softw 25(3):429–447
74. Chalmet L, Francis R, Saunders P (1982) Network models for building evacuation. Manag Sci 28:86–105
75. Cooke L, Halsey E (1966) The shortest route through a network with time-dependent internodal transit times. J Math Anal Appl 14:492–498
76. Ahuja RK, Orlin JB, Pallottino S, Scutella MG (2003) Dynamic shortest paths minimizing travel times and costs. Networks 41:197–205
77. Chabini L (1998) Discrete dynamic shortest path problems in transportation applications: complexity and algorithms with optimal run time. Transp Res Rec 1645:170–175
78. Orda A, Rom R (1995) On continuous network flows. Oper Res Lett 17:27–36
79. Pallottino S, Scutellà MG (1998) Shortest path algorithms in transportation models: classical and innovative aspects. In: Marcotte P, Nguyen S (eds) Equilibrium and advanced transportation modelling, Kluwer, pp 245–281
80. Hall A, Langkau K, Skutella M (2007) An FPTAS for quickest multicommodity flows with inflow-dependent transit times. Algorithmica 47:299–321
81. Fleisher LK, Scutella M (2007) Quickest flows over time. SIAM J Comput 36:1600–1630
82. Glockner GD, Nemhauser GL, Tovey CA (2001) Dynamic network flow with uncertain arc capacities: decomposition algorithm and computational results. Comput Optim Appl 18(3): 233–250

Chapter 2
Maximum and Minimum Cost Flow Finding in Networks in Fuzzy Conditions

The problems of the maximum and the minimum cost flow finding with zero and nonzero lower flow bounds are relevant, since they allow solving the problems of economic planning, logistics, transportation management, etc. In the area of transportation networks flow tasks enable to find the cargo transportation of the maximum volume between given points taking into account restrictions on the arc capacities of the cargo transmission paths, choose the routes of the optimal cost with the set lower flow bounds, which can be found after the profitability analysis of the cargo transportation along the particular road section. In considering these tasks it is necessary to take into account the inherent uncertainty of the network parameters, since environmental factors, measurement errors, repair work on the roads, the specifics of the constantly changing structure of the network influence the upper and lower flow bounds and transportation costs.

2.1 Maximum Flow Finding in a Network with Fuzzy Arc Capacities

The complex nature of environment factors, in particular, the inherent uncertainty is the basis for considering the maximum flow tasks in fuzzy dynamic network (Definition 2.1).

Definition 2.1 *Fuzzy transportation network is a fuzzy directed graph* [1, 2] $\widetilde{G} = (X, \widetilde{A})$, where $X = \{x_1, x_2, \ldots, x_n\}$ is the set of nodes, $\widetilde{A} = \{\mu_{\widetilde{A}}\langle x_i, x_j\rangle / \langle x_i, x_j\rangle\}$, $\langle x_i, x_j\rangle \in X^2$, $\mu_{\widetilde{A}}\langle x_i, x_j\rangle$—the fuzzy set of the arcs, where $\mu_{\widetilde{A}}(x_i, x_j)$ is a grade of membership of the directed arc $\langle x_i, x_j\rangle$ to the fuzzy set of the directed arcs $\widetilde{A}$. Fuzzy arc capacity is applied as $\mu_{\widetilde{A}}(x_i, x_j)$.

The key notion of the maximum flow task is the fuzzy residual (incremental) network, since we will search the maximum flow each time in the fuzzy residual

© Springer International Publishing Switzerland 2017

A.V. Bozhenyuk et al., *Flows in Networks Under Fuzzy Conditions*,
Studies in Fuzziness and Soft Computing 346, DOI 10.1007/978-3-319-41618-2_2

network that changes its structure based on the flow values passing along its arcs. Present the Definition 2.2 of the fuzzy residual network [3], modified for using in a fuzzy way.

Definition 2.2 *A fuzzy residual (incremental) transportation network* $\widetilde{G}^\mu$ *is a fuzzy network defined as a fuzzy directed graph that sets two arcs:* $\left(x_i^\mu, x_j^\mu\right)$ *with the residual arc capacity* $\tilde{u}_{ij}^\mu = \tilde{u}_{ij} - \tilde{\xi}_{ij}$ *and* $\left(x_j^\mu, x_i^\mu\right)$ *with the residual arc capacity* $\tilde{u}_{ji}^\mu = \tilde{\xi}_{ij}$ *for each arc* (x_i, x_j) *of the original graph* $\widetilde{G}$. *Thus, a fuzzy residual network contains only arcs with the positive arc capacities.*

Then the model of the maximum flow problem in a fuzzy transportation network [4] is defined as:

$$\tilde{v} = \sum_{x_j \in \Gamma(s)} \tilde{\xi}_{sj} = \sum_{x_k \in \Gamma^{-1}(t)} \tilde{\xi}_{kt} \to \max, \tag{2.1}$$

$$\sum_{x_j \in \Gamma(x_i)} \tilde{\xi}_{ij} = \sum_{x_k \in \Gamma^{-1}(x_i)} \tilde{\xi}_{ki} = \begin{cases} \tilde{v}, & x_i = s, \\ -\tilde{v}, & x_i = t, \\ \tilde{0}, & x_i \neq s, t, \end{cases} \tag{2.2}$$

$$\tilde{\xi}_{ij} \leq \tilde{u}_{ij}, \quad \forall \, (x_i, x_j) \in \widetilde{A}. \tag{2.3}$$

In the model (2.1)–(2.3) $\tilde{v}$—the maximum flow value in the fuzzy network; $\tilde{\xi}_{ij}$— fuzzy flow value passing along the arc (x_i, x_j); s—the initial node of the graph (the source); t—the terminal node of the graph (sink); $\Gamma(x_i)$—the set of nodes, arcs from the node $x_i \in X$ go to; $\Gamma^{-1}(x_i)$—the set of nodes, arcs from the node $x_i \in X$ go from; u_{ij}—the maximum amount of flow that can pass through the arc (x_i, x_j) (arc capacity).

It is necessary to perform the algorithm of solving the present task in fuzzy conditions. Let us introduce the method of the augmenting path finding in the network using breadth-first search [5] modified for using in fuzzy environment and for finding the shortest fuzzy path.

Method of the augmenting path finding by the breadth-first search in fuzzy conditions:

1. Form a queue consisting of the nodes Q. Initially Q contains only the source-node s.
2. Mark the node s as visited but without predecessor. Mark other nodes as unvisited.
3. Check the nodes in the queue:

   3.1 If the queue is empty, then stop and exit the algorithm, since there is no way.
   3.2 If the queue is not empty and the first node in the queue is $x_i = t$, go to the step 6.

    3.3 If the queue is not empty and the first node in the queue is $x_i \neq t$, delete the first node in the queue x_i.

4. Check the arcs $\{x_i, x_j\} \in \widetilde{A}$, such that the node x_j has not visited yet.

    4.1 If there are no such arcs $\{x_i, x_j\} \in \widetilde{A}$ that the node x_j has not visited yet, go to the step 3.

    4.2 If there are arcs $\{x_i, x_j\} \in \widetilde{A}$ that the node x_j has not visited yet, mark x_j as visited with the predecessor x_i.

5. Insert the node x_j at the end of the queue and go to the step 3.
6. Look through the nodes of the queue in the reverse order of t to s, each time passing to the predecessor. Return the path in the reverse order.

An algorithm for the maximum flow task solving in the network with fuzzy capacities

    Consider the algorithm for the maximum flow finding in the network with fuzzy capacities.

Step 1. Construct fuzzy residual network $\widetilde{G}^{\mu} = \left(X^{\mu}, \widetilde{A}^{\mu} \right)$, where X^{μ}—the set of nodes of the fuzzy residual network coincides with the set of nodes of the network $\widetilde{G}$ and $\widetilde{A}^{\mu} = \{ \langle \tilde{u}_{ij}^{\mu}/(x_i, x_j) \rangle \}$—fuzzy set of the arcs of the network $\widetilde{G}^{\mu}$ determined depending on the flow values coming along the arcs of the graph $\widetilde{G}$. Fuzzy arc capacities are used as the membership functions of the arcs. If $\tilde{\xi}_{ij} < \tilde{u}_{ij}$, then $\tilde{u}_{ij}^{\mu} = \tilde{u}_{ij}^{\mu} - \tilde{\xi}_{ij}$. If $\tilde{\xi}_{ij} > \tilde{0}$, then $\tilde{u}_{ji}^{\mu} = \tilde{\xi}_{ij}$. Initially the residual network coincides with the initial network (due to the equality of the arc flow to $\tilde{0}$).

Step 2. Search the shortest path $\widetilde{P}^{\mu}$ according to the criterion of the number of arcs from the source to the sink in the constructed fuzzy residual network, starting from zero flows. The selection is made using the breadth-first search in fuzzy conditions.

    2.1. If the path $\widetilde{P}^{\mu}$ is found, go to the **step 3**.

    2.2. If the path is failed to find, then the maximum flow $\tilde{\xi}_{ij} + \tilde{\delta}^{\mu} \widetilde{P}^{\mu} = \tilde{v}$ is obtained in the initial graph $\widetilde{G}$, the **exit**.

Step 3. Pass $\tilde{\delta}^{\mu} = \min \left[\tilde{u}_{ij}^{\mu} \right]$, $(x_i, x_j) \in \widetilde{P}^{\mu}$ flow units along the path.

Step 4. Update the flow values in $\widetilde{G}$: replace the flow $\tilde{\xi}_{ji}$ along the corresponding arcs (x_j, x_i) from $\widetilde{G}$ by $\tilde{\xi}_{ji} - \tilde{\delta}^{\mu}$ from $\tilde{\xi}_{ji}$ for arcs $\left(x_i^{\mu}, x_j^{\mu} \right) \notin \widetilde{A}$, $\left(x_i^{\mu}, x_j^{\mu} \right) \in \widetilde{A}^{\mu}$ in $\widetilde{G}^{\mu}$. Replace the flow $\tilde{\xi}_{ij}$ along the corresponding arcs (x_i, x_j) from $\widetilde{G}$ by $\tilde{\xi}_{ij} + \tilde{\delta}^{\mu}$ from $\tilde{\xi}_{ij}$ for arcs $\left(x_i^{\mu}, x_j^{\mu} \right) \in \widetilde{A}$, $\left(x_i^{\mu}, x_j^{\mu} \right) \in \widetilde{A}^{\mu}$ in $\widetilde{G}^{\mu}$. Replace the flow value in the graph $\widetilde{G}$: $\tilde{\xi}_{ij} \rightarrow \tilde{\xi}_{ij} + \tilde{\delta}^{\mu} \widetilde{P}^{\mu}$ and turn to the **step 1** starting with updated flow value along the arcs.

2.2 Method of Fuzzy Calculations with Fuzzy Numbers

It is necessary to perform sum, subtraction and comparing operations solving the maximum flow tasks and other tasks in the book in networks with fuzzy arc capacities. Conventional operations of the sum, subtraction are presented earlier in this chapter. Present methods can be used for solving flow tasks in fuzzy conditions, but they have some disadvantages.

The disadvantage of the conventional method of operating fuzzy numbers is strong "blurring" of borders of the resulting number and, as a result, loss of information content with such numbers. Standard subtraction operation of two equal fuzzy numbers does not lead to zero fuzzy number, which is unacceptable for flow tasks, where at least one arc must become saturated each iteration. It creates problems of the correct display of uncertainty of the proper system. In this case, we propose to use nonstandard subtraction operation [6], which does not lead to the strong "blurring" of the borders of the resulting number and produces a fuzzy number is equal to zero by subtracting equal fuzzy numbers. While specifying the fuzzy numbers the fact that the degree of borders blurring depends on the size of the center is not usually taken into account. Therefore, the more the center, the more "blurred" the borders should be (while measuring 1 kg of material, we are talking "about 1 kg", implying the number "from 900 to 1100 g", but while measuring 1 t. of material, imply that "about 1 t." is the number "from 990 to 1110 kg").

Otherwise, if absolute values are small, and deviation borders are large, the effectiveness of the fuzzy logic application decreases. Subtraction of fuzzy numbers in the case where it is impossible to uniquely identify the largest can lead to negative values, which is unacceptable for flows.

Let us represent two fuzzy triangular numbers: $\tilde{A}_1 = (10, 2, 3)$ and $\tilde{A}_2 = (9, 2, 7)$ of the form $\tilde{A}_1 = (a_1, \gamma_1, \delta_1)$ and $\tilde{A}_2 = (a_2, \gamma_2, \delta_2)$, where a_1 and a_2—the centers of triangular numbers, γ_1 and γ_2—the left deviations, δ_1 and δ_2—the right deviations. Assume, it is necessary to determine the subtraction of such numbers. It is necessary to determine the largest before subtraction of fuzzy numbers. We can't uniquely identify the largest number comparing fuzzy numbers $\tilde{A}_1$ и $\tilde{A}_2$, as $a_1 > a_2$, $a_1 - \gamma_1 > a_2 - \gamma_2$, $a_1 + \delta_1 < a_2 + \delta_2$, therefore, comparing them by the center of gravity, we find that the center of gravity of the second number is larger, therefore, it is estimated higher. Hence, the subtraction operation can be expressed as follows $(9,2,7)–(10,2,3) = (-1,5,9)$, i.e., we get fuzzy triangular number with a negative center and, accordingly, the negative left border, which contradicts the condition of non-negative flows values.

Therefore, a different approach to handling fuzzy numbers in the study of flows is required, which is in operating the central values of fuzzy numbers during the algorithm, and the operation of non-standard deduction will be used during the subtraction; "blurring" of the resulting number will be provided once at the end of the algorithm based on the basic values set by experts. It is considered that the numbers with the large centers have large uncertainty bounds.

Consequently, the following method [7] is proposed to use when operating triangular fuzzy numbers. Suppose there are the values of lower, upper flow bounds or transmission costs in a form of fuzzy triangular numbers on the number axis set by the expert according to his experience, knowledge and information about the particular road or its section. Then when adding two original triangular fuzzy numbers their centers will be added: $a_1 + a_2$, subtracted: $a_1 - a_2$, and $a_1 \geq a_2$. The fact that the number with the larger center will have the larger deviation borders is taken into account. To calculate the deviations it is necessary to define required value by adjacent values. Let the fuzzy parameter (for example, arc capacity) "near $.\tilde{a}'$." is between two adjacent values "near $\tilde{a}_1$" and "near $\tilde{a}_2$", $(\tilde{a}_1 \leq \tilde{a}' \leq \tilde{a}_2)$ which membership functions $\mu_{\tilde{a}_1}(a_1)$ and $\mu_{\tilde{a}_2}(a_2)$ have a triangular form, thus, the borders of membership function of fuzzy arc capacity $\mu_{\tilde{a}'}(a)$ of the fuzzy parameter "near $\tilde{x}$" one can set by the linear combination of left and right borders of adjacent values according to the method described in [8] and presented as:

$$l^L = \frac{(a_2 - a')}{(a_2 - a_1)} \times l_1^L + \left(1 - \frac{(a_2 - a')}{(a_2 - a_1)}\right) \times l_2^L,$$

$$l^R = \frac{(a_2 - a')}{(a_2 - a_1)} \times l_1^R + \left(1 - \frac{(a_2 - a')}{(a_2 - a_1)}\right) \times l_2^R. \tag{2.4}$$

In (2.4) l^L—the left deviation border of the fuzzy triangular number with the center a'; l^R—the right deviation border of the fuzzy triangular number with the center a'. It is shown in Fig. 2.1.

In the case when the central value of triangular number resulting by adding (subtracting) repeats the already marked value on the number axis, its deviation borders coincide with the deviation borders of the number marked on the number axis. If required central value is not between two numbers, but precedes the first marked value on the number axis, its deviation borders coincide with those of the first marked on the axis. The same applies to the case when the required central value follows the last marked value on the axis.

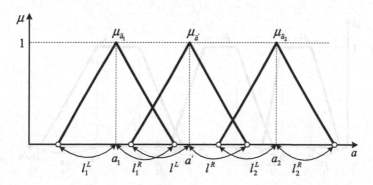

Fig. 2.1 Defining of the membership function $\mu_{\tilde{a}'}(a)$

The same method is used for calculations of the deviation borders of the trapezoidal numbers. Let us introduce the following equation for the left l^L and the right l^R trapezoidal deviation borders:

$$l^L = \frac{(a_2 - a')}{(a_2 - a_1)} \times l_1^L + \left(1 - \frac{(a_2 - a')}{(a_2 - a_1)}\right) \times l_2^L,$$

$$l^R = \frac{(b_2 - b')}{(b_2 - b_1)} \times l_1^R + \left(1 - \frac{(b_2 - b')}{(b_2 - b_1)}\right) \times l_2^R.$$

(2.5)

In the Eq. (2.5) l^L—the left deviation border of the fuzzy trapezoidal number with the center $[a', b']$; l^R—the right deviation border of the fuzzy trapezoidal number with the center $[a', b']$. It is represented in Fig. 2.2.

The present method allow us to avoid receiving of the negative fuzzy numbers, which are unacceptable for the flows, simplifies the calculations and leads to the strong "blurring" of the borders of fuzzy numbers.

Let us consider the numerical example that implements the algorithm for the maximum flow finding in the network with fuzzy arc capacities.

Numerical example 1
Let that network is presented in the form of the fuzzy directed graph, given in Fig. 2.3.

The arc capacities in the form of the fuzzy intervals set by the experts are assigned to the arcs of the graph. It is necessary to find the maximum flow in the graph and perform it in the form of the fuzzy trapezoidal number, if the basic values of arc capacities in the form of the fuzzy trapezoidal numbers are given, as shown in Fig. 2.4.

Step 1. Fuzzy residual network coincides with the initial network $\widetilde{G}$, shown in Fig. 2.3 by the equality of the arc flows to $[\tilde{0}, \tilde{0}]$.

Step 2. Search the shortest path according to the number of arcs from s to t in fuzzy residual network. Use the breadth-first-search for it:

The queue Q consists of the vertices s. Note s as visited without predecessors.

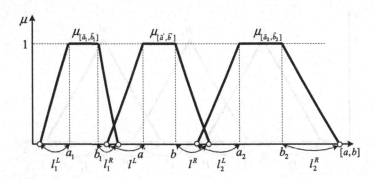

Fig. 2.2 Defining of the membership function $\mu_{\tilde{a}'}(a)$

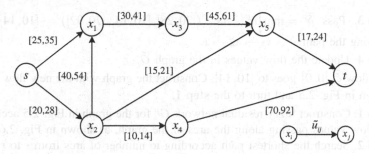

Fig. 2.3 The initial network $\widetilde{G}$

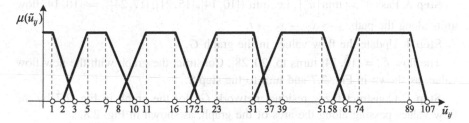

Fig. 2.4 Membership functions of the basic values of arc capacities of the network $\widetilde{G}$

Delete the first node in queue s. Form the set of nodes arcs from s go to and that have not yet visited: $\{x_1, x_2\}$. The queue consists of the nodes $\{x_1, x_2\}$. The nodes s, x_1, x_2 have been visited. The nodes x_1, x_2 have the predecessor s.

Delete the first node in queue x_1. Form the set of nodes arcs from x_1 go to and that have not yet visited: $\{x_3\}$. The queue consists of the nodes $\{x_2, x_3\}$. The nodes s, x_1, x_2, x_3 have been visited. The nodes x_1, x_2 have the predecessor s, the node x_3 has the predecessor x_1.

Delete the first node in queue x_2. Form the set of nodes arcs from x_2 go to and that have not yet visited: $\{x_4, x_5\}$. The queue consists of the nodes $\{x_3, x_4, x_5\}$. The nodes $s, x_1, x_2, x_3, x_4, x_5$ have been visited. The nodes x_1, x_2 have the predecessor s, the node x_3 has the predecessor x_1, the nodes x_4, x_5 have the predecessor x_2.

Delete the first node in queue x_3. Form the set of nodes arcs from x_3 go to and that have not yet visited: $\{\emptyset\}$. The node x_3 is delete from the queue. The queue consists of the nodes $\{x_4, x_5\}$. The nodes $s, x_1, x_2, x_3, x_4, x_5$ have been visited. The nodes x_1, x_2 have the predecessor s, the node x_3 has the predecessor x_1, the nodes x_4, x_5 have the predecessor x_2.

Delete the first node in the queue x_4. Form the set of nodes arcs from x_4 go to and that have not yet visited: $\{t\}$. The arc (x_4, t) is found and the algorithm terminates. The queue consists of the nodes $\{x_5, t\}$. All nodes have been visited. The nodes x_1, x_2 have the predecessor s, the node x_3 has the predecessor x_1, the nodes x_4, x_5 have the predecessor x_2, the node t has the predecessor x_4. Go backwards by predecessors and receive the path $s \rightarrow x_2 \rightarrow x_4 \rightarrow t$.

Step 3. Pass $\tilde{\delta}^{\mu} = \min\left[\tilde{u}^{\mu}_{ij}\right]$, i.e. $([2\tilde{0}, 2\tilde{8}], [1\tilde{0}, 1\tilde{4}], [7\tilde{0}, 9\tilde{2}]) = [1\tilde{0}, 1\tilde{4}]$ flow units along the path $s \rightarrow x_2 \rightarrow x_4 \rightarrow t$.

Step 4. Update the flow values in the graph $\widetilde{G}$.

The flow $\tilde{\xi}_{ij}[\tilde{0}, \tilde{0}]$ goes to $[1\tilde{0}, 1\tilde{4}]$. Construct the graph with the new flow value, as shown in Fig. 2.5 and turn to the **step 1**.

Step 1. Construct fuzzy residual network $\widetilde{G}^{\mu}$ for the graph in Fig. 2.5 according to the flow values passing along the arcs of the graph, as shown in Fig. 2.6.

Step 2. Search the shortest path according to number of arcs from s to t in the fuzzy residual network $\widetilde{G}^{\mu}$. Use the breadth-first-search and find the path: $s \rightarrow x_2 \rightarrow x_5 \rightarrow t$.

Step 3. Pass $\tilde{\delta}^{\mu} = \min\left[\tilde{u}^{\mu}_{ij}\right]$, i.e. $\min\left([1\tilde{0}, 1\tilde{4}], [1\tilde{5}, 2\tilde{1}], [1\tilde{7}, 2\tilde{4}]\right) = [1\tilde{0}, 1\tilde{4}]$ flow units along the path: $s \rightarrow x_2 \rightarrow x_5 \rightarrow t$.

Step 4. Update the flow values in the graph $\widetilde{G}$.

The flow $\tilde{\xi}_{ij} = [1\tilde{0}, 1\tilde{4}]$ turns to $[2\tilde{0}, 2\tilde{8}]$. Construct the graph with the new flow value, as shown in Fig. 2.7 and turn to the **step 1**.

Step 1. Construct fuzzy residual network $\widetilde{G}^{\mu}$ for the graph in Fig. 2.7 by the flow values passing along the arcs of the graph, as shown in Fig. 2.8.

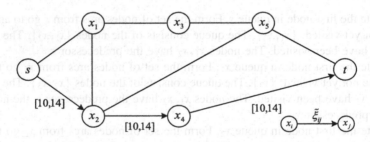

Fig. 2.5 Graph $\widetilde{G}$ with the new flow value of [10,14] units

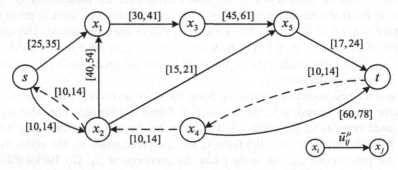

Fig. 2.6 Fuzzy residual network $\widetilde{G}^{\mu}$ for the graph in Fig. 2.5

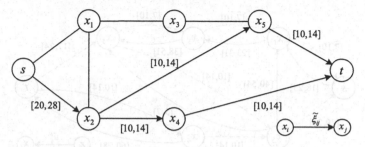

Fig. 2.7 $\widetilde{G}$ with the new flow value of $[2\tilde{0}, 2\tilde{8}]$ units

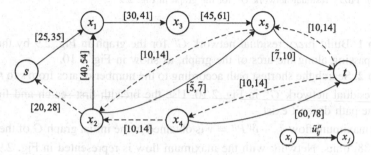

Fig. 2.8 Fuzzy residual network $\widetilde{G}^{\mu}$ for the graph in Fig. 2.7

Step 2. Search the shortest path according to number of arcs from s to t in the fuzzy residual network $\widetilde{G}^{\mu}$ in Fig. 2.8. Use the breadth-first-search and find the path: $s \rightarrow x_1 \rightarrow x_3 \rightarrow x_5 \rightarrow t$.

Step 3. Pass $\tilde{\delta}^{\mu} = \min\left[\tilde{u}_{ij}^{\mu}\right]$, i.e. min $([2\tilde{5}, 3\tilde{5}], [3\tilde{0}, 4\tilde{1}], [4\tilde{5}, 6\tilde{1}], [\tilde{7}, 1\tilde{0}]) = [\tilde{7}, 1\tilde{0}]$ flow units along the path $s \rightarrow x_1 \rightarrow x_3 \rightarrow x_5 \rightarrow t$.

Step 4. Update the flow values in the graph $\widetilde{G}$.

The flow $\tilde{\xi}_{ij} = [2\tilde{0}, 2\tilde{8}]$ turns to $[2\tilde{0}, 2\tilde{8}] + [\tilde{7}, 1\tilde{0}] = [2\tilde{7}, 3\tilde{8}]$. Construct the graph with the new flow value, as shown in Fig. 2.9 and turn to the **step 1**.

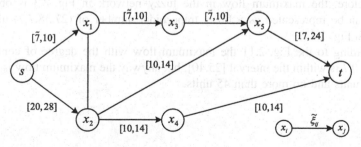

Fig. 2.9 Graph $\widetilde{G}$ with the new flow value of $[27,38]$ units

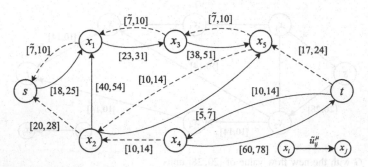

Fig. 2.10 Fuzzy residual network $\widetilde{G}^{\mu}$ for the graph in Fig. 2.9

Step 1. Build fuzzy residual network $\widetilde{G}^{\mu}$ for the graph in Fig. 2.9 by the flow values passing along the arcs of the graph, as show in Fig. 2.10.

Step 2. Search the shortest path according to the number of arcs from s to t in the fuzzy residual network $\widetilde{G}^{\mu}$ in Fig. 2.10. Use the breadth-first-search and find the path: the path doesn't exist.

The maximum flow $\tilde{\xi}_{ij} + \tilde{\delta}^{\mu}\widetilde{P}^{\mu} = \tilde{v}$ is obtained in the initial graph $\widetilde{G}$ of the value of $[2\tilde{7}, 3\tilde{8}]$ units. Network with the maximum flow is represented in Fig. 2.9.

Let us define deviation borders of the obtained fuzzy interval $[2\tilde{7}, 3\tilde{8}]$ corresponded to the maximum flow in the graph $\widetilde{G}$. The detected result is between two adjacent basic values of the arc capacities: $[2\tilde{3}, 3\tilde{1}]$ with the left deviation $l_1^L = 6$, the right deviation—$l_1^R = 6$ and $[3\tilde{9}, 5\tilde{1}]$ with the left deviation $l_2^L = 8$, the right deviation $-l_2^R = 10$. According to (2.5) we obtain:

$$l^L = \frac{(a_2 - a')}{(a_2 - a_1)} \times l_1^L + \left(1 - \frac{(a_2 - a')}{(a_2 - a_1)}\right) \times l_2^L = \frac{(39 - 27)}{(39 - 23)} \times 6 + \left(1 - \frac{(39 - 27)}{(39 - 23)}\right) \times 8$$
$$= 6.5 \approx 7,$$

$$l^R = \frac{(b_2 - b')}{(b_2 - b_1)} \times l_1^R + \left(1 - \frac{(b_2 - b')}{(b_2 - b_1)}\right) \times l_2^R = \frac{(51 - 37)}{(51 - 31)} \times 6 + \left(1 - \frac{(51 - 37)}{(51 - 31)}\right) \times 10$$
$$= 7.2 \approx 7.$$

Therefore, the maximum flow in the fuzzy network in Fig. 2.3 is obtained, which can be represented as a fuzzy trapezoidal number of (27,38,7,7) units, as shown in Fig. 2.11.

According to the Fig. 2.11 the maximum flow with the degree of confidence equal to 0.7 is within the interval [25,40], but anyway the maximum flow is no less than 20 units and no more than 45 units.

Fig. 2.11 The maximum
flow in the form of the fuzzy
trapezoidal number of
(27,38,7,7) units

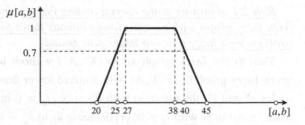

2.3 Maximum Flow Finding in a Network with Fuzzy Nonzero Lower and Upper Flow Bounds

Let the network considered in the previous section contains the lower flow bounds besides arc capacities. It occurs in the case, when one deals with transportation profitability. For example, a transport plane will carry out the flight at the lowest feasible amount of load of 1 ton; passenger aircraft will fly if at least twenty tickets will be sold. These figures are considered as lower flow bounds. Thus, nonzero lower flow bounds must be considered in some flow problems. The capacity in such problems is the upper flow bound. Thus, the flow must satisfy the constraint on upper and lower flow bounds, i.e. expression (1.13).

The complexity of the problem statement of the maximum flow with lower bounds specified on the arcs (1.11)–(1.13) lies in the fact that the feasible flow cannot exist, in contrast to the conventional problem of the maximum flow finding (zero flow is valid). Therefore it is necessary to solve two problems: on the admissibility of the flow, and then—to find the maximum flow in the case of the existence of the feasible flow. Considering the fuzzy nature of the upper and lower flow bounds, we come to the maximum flow problem formulation in the network, taking into account the nonzero lower and upper flow bounds, presented in the fuzzy form [9]:

$$\tilde{v} = \sum_{x_j \in \Gamma(s)} \tilde{\xi}_{sj} = \sum_{x_k \in \Gamma^{-1}(t)} \tilde{\xi}_{kt} \to \max, \tag{2.6}$$

$$\sum_{x_j \in \Gamma(x_i)} \tilde{\xi}_{ij} = \sum_{x_k \in \Gamma^{-1}(x_i)} \tilde{\xi}_{ki} = \begin{cases} \tilde{v}, & x_i = s, \\ -\tilde{v}, & x_i = t, \\ \tilde{0}, & x_i \neq s, t, \end{cases} \tag{2.7}$$

$$\tilde{l}_{ij} \leq \tilde{\xi}_{ij} \leq \tilde{u}_{ij}, \quad \forall \, (x_i, x_j) \in \tilde{A}. \tag{2.8}$$

In the model (2.6)–(2.8) $\tilde{l}_{ij}$—fuzzy lower flow bound for the arc (x_i, x_j).

To solve this problem one must turn to the fuzzy graph without lower bounds flows [10], that is reflected in the introduced *rule 2.1* of turning to the corresponding fuzzy graph without lower flow bounds from the fuzzy graph with nonzero lower flow bounds, modified for the using in the fuzzy form:

Rule 2.1 of turning to the corresponding fuzzy graph without lower flow bounds from fuzzy graph with nonzero lower bounds flows for solving the maximum flow problem with fuzzy nonzero lower flow bounds

Turn to the fuzzy graph $\widetilde{G}^* = (X^*, \widetilde{A}^*)$ without lower flow bounds from the given fuzzy graph $\widetilde{G} = (X, \widetilde{A})$ with nonzero lower flow bounds. Introduce artificial nodes s^* and t^*, the arc (t, s) with $\tilde{u}_{ts}^* = \infty$, $\tilde{l}_{ts}^* = \tilde{0}$ in the new graph $\widetilde{G}$. For each node (x_i, x_j) in $\widetilde{G}$ with $\tilde{l}_{ij} \neq \tilde{0}$: (1) decrease $\tilde{u}_{ij}$ to $\tilde{u}_{ij}^* = \tilde{u}_{ij} - \tilde{l}_{ij}$, $\tilde{l}_{ij}$ to $\tilde{0}$. (2) Introduce arcs (s^*, x_j) and (x_i, t^*) with flow bounds equal to $\tilde{u}_{s^* x_j}^* = \tilde{u}_{x_i t^*}^* = \tilde{l}_{ij}$, $\tilde{l}_{s^* x_j}^* = \tilde{l}_{x_i t^*}^* = \tilde{0}$. Arcs without lower flow bounds are the same for $\widetilde{G}^*$: for any arc (x_i, x_j) with $\tilde{l}_{ij} = \tilde{0}$ is $\tilde{u}_{ij}^* = \tilde{u}_{ij}$.

Thus, artificial arcs have capacities equal to the lower flow bounds in the generated graph without lower flow bounds and initial arcs with nonzero lower bounds are introduced with a capacity equals the difference between the initial upper and lower flow bound. Artificial arc from t and s defines the feasible flow in the original graph.

Start to search for the maximum flow in the fuzzy residual network corresponding to the graph without lower flow bounds [10] and constructed according to the Definition 2.3, modified for use in a fuzzy way after the transition to the corresponding graph without lower flow bounds from the fuzzy graph. The criterion of the feasible flow existing in the initial fuzzy graph is finding the maximum flow equals the sum of the lower flow bounds in the transformed graph without lower bounds.

Definition 2.3 *Fuzzy residual network for the graph* $\widetilde{G}^* = \left(X^*, \widetilde{A}^*\right)$ *without lower flow bounds for solving the maximum flow problem with fuzzy nonzero lower flow bounds—the network* $\widetilde{G}^{*\mu} = \left(X^{*\mu}, \widetilde{A}^{*\mu}\right)$, *where* $X^{*\mu} = X^*$—*the set of the nodes of the fuzzy residual network with artificial nodes,* $\widetilde{A}^{*\mu} = \left\{\left\langle \tilde{u}_{ij}^{*\mu} / \left(x_i^{*\mu}, x_j^{*\mu}\right)\right\rangle\right\}$—*fuzzy set of the arcs of the network* $\widetilde{G}^{*\mu}$, *constructed according the following rules: for all arcs, if* $\tilde{\xi}_{ij}^* < \tilde{u}_{ij}^*$, *then include corresponding arc in* $\widetilde{G}^{*\mu}$ *with arc capacity* $\tilde{u}_{ij}^{*\mu} = \tilde{u}_{ij}^* - \tilde{\xi}_{ij}^*$. *For all arcs, if* $\tilde{\xi}_{ij}^* > \tilde{0}$, *then include corresponding arc in* $\widetilde{G}^{*\mu}$ *with arc capacity* $\tilde{u}_{ji}^{*\mu} = \tilde{\xi}_{ij}^*$.

Reverse transition to the initial graph with the feasible flow from the modified graph with the maximum flow [11] is according to the introduced *rule 2.2*, modified for the use in fuzzy terms.

The rule 2.2 of transition to the graph with the feasible flow from the graph without lower flow bounds with the maximum flow for solving the maximum flow problem with fuzzy nonzero lower flow bounds

Turn to the graph $\widetilde{G}$ from the graph $\widetilde{G}^*$ as following: reject artificial nodes and arcs, connecting them with other nodes. The feasible flow vector $\tilde{\xi} = \left(\tilde{\xi}_{ij}\right)$ of the value $\tilde{\sigma}$ is defined as: $\tilde{\xi}_{ij} = \tilde{\xi}_{ij}^* + \tilde{l}_{ij}$, where $\tilde{\xi}_{ij}^*$—the flows, going along the arcs of the graph $\widetilde{G}^*$ after deleting all artificial nodes and connecting arcs.

If the feasible flow is found in the initial fuzzy graph, transform it by finding the maximum [10], according to the *rule 2.3*, presented in the fuzzy form. If the maximum flow equals the sum of the lower flow bounds *is not found in the modified graph, therefore, there is no feasible flow in the fuzzy initial graph and the task has no solution*

The rule 2.3 of the fuzzy residual network constructing with the feasible flow vector for solving the maximum flow problem with fuzzy nonzero lower flow bounds

For all arc, if $\tilde{\xi}_{ij} < \tilde{u}_{ij}$, then include the corresponding arc in $\widetilde{G}^\mu\left(\tilde{\xi}\right)$ with arc capacity $\tilde{u}_{ij}^\mu = \tilde{u}_{ij} - \tilde{\xi}_{ij}$. For all arc, if $\tilde{\xi}_{ij} > \tilde{l}_{ij}$, then include the corresponding arc in $\widetilde{G}^\mu\left(\tilde{\xi}\right)$ with arc capacity $\tilde{u}_{ji}^\mu = \tilde{\xi}_{ij} - \tilde{l}_{ij}$.

Let us represent the formal algorithm for solving the proposed problem based on the defined rules and definitions.

Algorithm of the maximum flow finding in the network with nonzero lower flow bounds in fuzzy conditions

Let us represent the formal algorithm for solving the proposed problem based on the defined rules and definitions [9].

Step 1. Let us define if the initial graph $\widetilde{G} = \left(X, \widetilde{A}\right)$ has the feasible flow. Turn to the graph $\widetilde{G}^* = \left(X^*, \widetilde{A}^*\right)$ without lower flow bounds according to the *rule 2.1*.

Step 2. Find maximum flow in $\widetilde{G}^*$ between artificial nodes. Build a fuzzy residual network starting with zero flows according to the Definition 2.3.

Step 3. Search the shortest path $\widetilde{P}^{*\mu}$ in terms of the number of arcs from the artificial source s^* to the artificial sink t^* in the constructed fuzzy residual network starting with zero flow values. The choice of the shortest path is according to the breadth-first search.

3.1. If the $\widetilde{P}^{*\mu}$ is found, go to the **step 4**.

3.2. The flow value $\tilde{\phi} < \sum_{\tilde{l}_{ij} \neq 0} \tilde{l}_{ij}$ is obtained, which is the maximum flow in $\widetilde{G}^*$, if the path is failed to find. It means that it is impossible to pass any unit of flow, but not all the artificial arcs are saturated. Therefore, initial graph $\widetilde{G}$ has no feasible flow and the task has no solution. Exit.

Step 4. Pass the minimum from the arc capacities $\tilde{\delta}^{*\mu} = \min\left[\tilde{u}\left(\widetilde{P}^{*\mu}\right)\right]$, $\tilde{u}\left(\widetilde{P}^{*\mu}\right) = \min\left[\tilde{u}_{ij}^{*\mu}\right]$, $\left(x_i^{*\mu}, x_j^{*\mu}\right) \in \widetilde{P}^{*\mu}$ along the path $\widetilde{P}_p^{*\mu}$.

Step 5. Update the fuzzy flow values in the graph $\widetilde{G}^*$: change the fuzzy flow $\tilde{\xi}_{ji}^*$ along the corresponding arcs $\left(x_j^*, x_i^*\right)$ from $\widetilde{G}^*$ by $\tilde{\xi}_{ji}^* - \tilde{\delta}^{*\mu}$ for arcs $\left(x_i^{*\mu}, x_j^{*\mu}\right) \notin \widetilde{A}^*$, $\left(x_i^{*\mu}, x_j^{*\mu}\right) \in \widetilde{A}^{*\mu}$ in $\widetilde{G}^{*\mu}$. Change the fuzzy flow $\tilde{\xi}_{ij}^*$ along the arcs

$\left(x_i^*, x_j^*\right)$ from $\widetilde{G}^*$ by $\widetilde{\xi}_{ij}^* + \widetilde{\delta}^{*\mu}$ for arcs $\left(x_i^{*\mu}, x_j^{*\mu}\right) \in \widetilde{A}^*$, $\left(x_i^{*\mu}, x_j^{\mu*}\right) \in \widetilde{A}^{\mu*}$ in $\widetilde{G}^{*\mu}$.
Replace $\widetilde{\xi}_{ij}^*$ by $\widetilde{\xi}_{ij}^* + \widetilde{\delta}^{*\mu}\widetilde{P}^{*\mu}$.

Step 6. Provide a comparison of the flow vector $\widetilde{\xi}_{ij}^* + \widetilde{\delta}^{*\mu} \times P^{*\mu}$ of the value $\widetilde{\sigma}^*$
and the sum of the lower flow bounds $\sum_{\widetilde{l}_{ij} \neq \widetilde{0}} \widetilde{l}_{ij}$ of the original graph:

6.1. If the flow vector $\widetilde{\xi}_{ij}^* + \widetilde{\delta}^{*\mu} \times \widetilde{P}^{*\mu}$ of the value $\widetilde{\sigma}^*$ is less than $\sum_{\widetilde{l}_{ij} \neq \widetilde{0}} \widetilde{l}_{ij}$, i.e.
not all artificial arcs become saturated, go to the **step 2**, i.e. to constructing the
new incremental graph with the flow passing along the arcs until $\widetilde{\xi}_{ij}^* + \widetilde{\delta}^{*\mu}\widetilde{P}^{*\mu}$
becomes equal to $\sum_{\widetilde{l}_{ij} \neq \widetilde{0}} \widetilde{l}_{ij}$.

6.2. If the flow value $\widetilde{\xi}_{ij}^* + \widetilde{\delta}^{*\mu} \times P^{*\mu}$ is equal to $\sum_{\widetilde{l}_{ij} \neq \widetilde{0}} \widetilde{l}_{ij}$, i.e. all arcs from the
artificial source to the artificial sink become saturated, then the value $\widetilde{\xi}_{ij}^* + \widetilde{\delta}^{*\mu} \times P^{*\mu}$
is required value of maximum flow $\widetilde{\sigma}^*$ in $\widetilde{G}^*$ In this case the flow $\widetilde{\xi}_{ts}^*$ passing along
the artificial arc (t, s) in $\widetilde{G}^*$ determines the feasible flow in the initial graph $\widetilde{G}$ of the
value $\widetilde{\sigma} = \widetilde{\xi}_{ts}^*$. Turn to the graph $\widetilde{G}$ from the graph $\widetilde{G}^*$ according to the *rule 2.2*. The
network $G\left(\widetilde{\xi}\right)$ is obtained. Go to the **step 7**.

Step 7. Construct the residual network $G\left(\widetilde{\xi}^{\mu}\right)$ taking into account the feasible
flow vector $\widetilde{\xi} = \left(\widetilde{\xi}_{ij}\right)$ in the graph $\widetilde{G}$ according to the *rule 2.3*.

Step 8. Define the shortest path $\widetilde{P}^{\mu}$ according to the number of arcs from the
artificial source to the artificial sink in the constructed residual network $G\left(\widetilde{\xi}^{\mu}\right)$. The
choice of the shortest path is according to the breadth-first search.

8.1. Go to the **step 9** if the augmenting path $\widetilde{P}^{\mu}$ is found.

8.2. The maximum flow $\widetilde{\xi}_{ij} + \widetilde{\delta}^{\mu} \times \widetilde{P}^{\mu} = \widetilde{v}$ in $\widetilde{G}$ is found if the path is failed to
find, then stop.

Step 9. Pass $\widetilde{\delta}^{\mu} = \min\left[\widetilde{u}\left(\widetilde{P}^{\mu}\right)\right]$, $\widetilde{u}\left(\widetilde{P}^{\mu}\right) = \min\left[\widetilde{u}_{ij}^{\mu}\right]$, $\left(x_i^{\mu}, x_j^{\mu}\right) \in \widetilde{P}^{\mu}$ along the
found path.

Step 10. Update the fuzzy flow values in the graph $\widetilde{G}$: replace the fuzzy flow $\widetilde{\xi}_{ji}$
along the corresponding arcs (x_j, x_i) from $\widetilde{G}$ by $\widetilde{\xi}_{ji} - \widetilde{\delta}^{\mu}$ for arcs $\left(x_i^{\mu}, x_j^{\mu}\right) \notin \widetilde{A}$,
$\left(x_i^{\mu}, x_j^{\mu}\right) \in \widetilde{A}^{\mu}$ in $\widetilde{G}^{\mu}\left(\widetilde{\xi}\right)$ and change the fuzzy flow $\widetilde{\xi}_{ij}$ along the arcs (x_i, x_j) from
$\widetilde{G}$ by $\widetilde{\xi}_{ij} + \widetilde{\delta}^{\mu}$ for arcs $\left(x_i^{\mu}, x_j^{\mu}\right) \in \widetilde{A}$, $\left(x_i^{\mu}, x_j^{\mu}\right) \in \widetilde{A}^{\mu}$, in $\widetilde{G}^{\mu}\left(\widetilde{\xi}\right)$ and replace the flow
value in $\widetilde{G}$: $\xi_{ij} \rightarrow \widetilde{\xi}_{ij} + \widetilde{\delta}^{\mu} \times \widetilde{P}^{\mu}$ and turn to the **step 7** starting from the new flow
value along the arcs.

Thus the described algorithm allows to find the maximum flow in networks with
lower flow bounds in fuzzy conditions or show that the feasible flow doesn't exist.

Let us consider the proof of the main provisions of the proposed algorithm.

Let us show that if there is the maximum flow equals the sum of the lower flow bounds in the graph $\widetilde{G}^*$, there is the feasible flow the original graph. It is necessary to introduce the Theorem 2.1.

Theorem 2.1 *If the maximum flow in the graph $\widetilde{G}^*$ is equal to the sum of the lower flow bounds $\tilde{\sigma}^* = \sum_{\tilde{l}_{ij} \neq \tilde{0}} \tilde{l}_{ij}$, therefore, there is the feasible flow of the value $\tilde{\sigma} = \tilde{\xi}^*_{ts}$ in the original graph $\widetilde{G}$.*

Proof Let us assume that it is false, i.e. if the maximum flow in the graph $\widetilde{G}^*$ is not equal to the sum of the lower flow bounds $\tilde{\sigma}^* \neq \sum_{\tilde{l}_{ij} \neq \tilde{0}} \tilde{l}_{ij}$, therefore, there is the feasible flow of the value $\tilde{\sigma} = \tilde{\xi}^*_{ts}$ in the graph $\widetilde{G}$. Graph $\widetilde{G}^*$ is obtained by addition of the parameter $\tilde{l}_{ij}$ to artificial arcs and subtraction of the parameter $\tilde{l}_{ij}$ from $\tilde{u}_{ij}$ from the arcs with nonzero lower flow bounds. The flow passing along the arc (t, s) defines the feasible one. Applying the inverse transformation assume that the flow is not equal to the sum of the lower flow bounds in $\widetilde{G}^*$ gives the feasible one in $\widetilde{G}$. Since the flow leaving the artificial arcs of the graph $\widetilde{G}^*$ can not be than $\sum_{\tilde{l}_{ij} \neq \tilde{0}} \tilde{l}_{ij}$, consider the case when the flow $\tilde{\sigma}^* < \sum_{\tilde{l}_{ij} \neq \tilde{0}} \tilde{l}_{ij}$ Then, by subtracting $\tilde{l}_{ij}$ from the flow $\tilde{\xi}^*_{ij}$ along the artificial arcs of the graph $\widetilde{G}^*$, obtain negative flow values (because the flow less than $\tilde{l}_{ij}$ is passing the artificial arcs), which contradicts the non-negativity flow condition as well as the fact that during equivalent transformations we had to get zero values of the flow and liquidate artificial arcs. Therefore, our hypothesis is not true, and the theorem is proved.

Let us show that if the feasible flow in $\widetilde{G}^*$ is equal to the sum of the lower flow bounds in the original graph, and it has a value of $\tilde{\sigma} = \tilde{\xi}^*_{ts}$ and defined as $\tilde{\xi}_{ij} = \tilde{\xi}^*_{ij} + \tilde{l}_{ij}$. It is necessary to introduce the Corollary 2.1.

Corollary 2.1 *If $\tilde{\sigma}^* = \sum_{\tilde{l}_{ij} \neq \tilde{0}} \tilde{l}_{ij}$, than the feasible flow vector $\tilde{\xi} = \left(\tilde{\xi}_{ij} \right)$ of the value $\tilde{\sigma}$ and defining as $\tilde{\xi}_{ij} = \tilde{\xi}^*_{ij} + \tilde{l}_{ij}$, is the feasible flow in $\widetilde{G}$ of the value $\tilde{\sigma} = \tilde{\xi}^*_{ts}$.*

Proof The proof follows from the Theorem 2.1 and the rules of transition to the graph without lower flow bounds $\widetilde{G}^*$. Since the lower flow bounds are deducted from the arc capacities during transition from the graph $\widetilde{G}$ to the graph $\widetilde{G}^*$, then artificial arcs are rejected and lower flow bounds are added to the flow value when reverse transition to the graph $\widetilde{G}$. Thus, we subtract and then add the same amount. The following flow is the feasible one because the addition of lower flow bounds ensures that the flow greater or equal to the lower flow bound of this arc will pass along any arc of the graph, i.e. conditions on the flow restrictions are satisfied for any arc of the graph $\widetilde{G}$. The corollary is proved.

Let us show that the flow defined in a fuzzy form and received on the **step 8** of the maximum flow algorithm with nonzero lower flow bounds is the fuzzy maximum flow in the original graph $\widetilde{G}$ based on the Theorem 2.2.

Theorem 2.2 *Fuzzy flow* $\tilde{\xi}_{ij} + \tilde{\delta}^\mu \times \widetilde{P}^\mu = \tilde{v}$ *obtained at the* **step 8** *of the maximum flow algorithm with nonzero lower flow bounds is a fuzzy maximum flow in the original graph* $\widetilde{G}$.

Proof To show that the fuzzy flow $\tilde{\xi}_{ij} + \tilde{\delta}^\mu \times \widetilde{P}^\mu = \tilde{v}$ is the maximum flow in the graph $\widetilde{G}$, assume that it is not true. Then the incremental fuzzy path from the source to the sink must exit in the graph $\widetilde{G}$. However, the algorithm terminates when the path doesn't exist, therefore, we have a contradiction. Thus, the obtained flow $\tilde{\xi}_{ij} + \tilde{\delta}^\mu \times \widetilde{P}^\mu = \tilde{v}$ is the maximum flow in a graph. The theorem is proved.

Let us show, if the maximum flow in the graph $\widetilde{G}^*$ is less than the sum of the lower flow bounds of the initial graph, then the feasible flow doesn't exist in $\widetilde{G}$, that is reflected in the Corollary 2.2.

Corollary 2.2 *If the maximum flow* $\tilde{\sigma}^*$ *in* $\widetilde{G}^*$ *is less than the sum of the lower flow bounds of the initial graph, i.e.* $\tilde{\sigma}^* < \sum_{\tilde{l}_{ij} \neq \tilde{0}} \tilde{l}_{ij}$, *then the feasible flow doesn't exist in* $\widetilde{G}$.

Proof At the reverse transition to the initial graph from the transformed graph without lower flow bounds the flow value $\tilde{\sigma}^* < \sum_{\tilde{l}_{ij} \neq \tilde{0}} \tilde{l}_{ij}$ means that the maximum flow in the original graph is less than the sum of the lower flow bounds, i.e. for any arc which has nonzero lower flow bound the flow greater than or equal to it can not be transferred. Hence, the feasible flow doesn't exist. The corollary is proved.

Consider the numerical example, which implements operations of the maximum flow finding algorithm in fuzzy network with nonzero lower flow bounds.

Numerical example 2
Let the network is represented in the form of the fuzzy directed graph given in Fig. 2.12. Values of the upper and lower flow bounds in the form of fuzzy intervals are assigned to the arcs of the graph. It is necessary to find the maximum flow in the graph and represent it in the form of the fuzzy triangular number, if the basic values of arc capacities in the form of trapezoidal fuzzy numbers are known, as shown in Fig. 2.4.

Fig. 2.12 Initial network $\widetilde{G}$

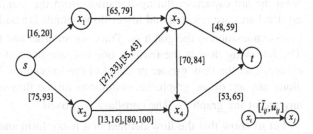

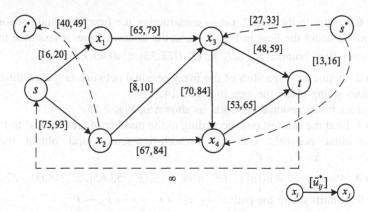

Fig. 2.13 Graph $\widetilde{G}^*$ without lower flow bounds

Step 1. Turn to the graph without lower flow bounds according to the *rule 2.1*, as shown in Fig. 2.13.

Step 2. Fuzzy residual network $\widetilde{G}^{*\mu}$ at the **step 2** coincides with the graph $\widetilde{G}^*$ without lower flow bounds, presented in Fig. 2.13 by the equality of the arc flows to $[\tilde{0}, \tilde{0}]$.

Step 3. Find the shortest path according to the number of arcs from s^* to t^* in the fuzzy residual network $\widetilde{G}^{*\mu}$ in Fig. 2.13 by the breadth-first-search. Obtain the path $s^* \to x_3 \to t \to s \to x_2 \to t^*$.

Step 4. Pass $\tilde{\delta}^{*\mu} = \min\left[\tilde{u}_{ij}^{\mu}\right]$, i.e. min $([27,33], [48,59], \infty, [75,93], [40,49]) = [27,33]$ flow units along the path $s^* \to x_3 \to t \to s \to x_2 \to t^*$.

Step 5. Update the values of flows in $\widetilde{G}^*$.

The flow $\tilde{\xi}_{ij}^* = [\tilde{0}, \tilde{0}]$ turns to $\tilde{\xi}_{ij}^* = [\tilde{0}, \tilde{0}] + [27,33] = [27,33]$.

Construct a graph with the new flow value, as shown in Fig. 2.14.

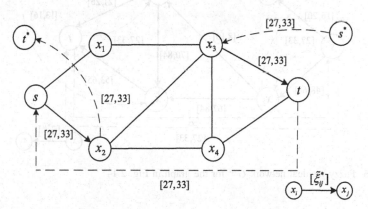

Fig. 2.14 Graph $\widetilde{G}^*$ with the new flow value of [27,33] units

Step 6.1. Turn to the **step 2**, i.e. to constructing the fuzzy residual network $\widetilde{G}^{*\mu}$ taking into account the flow in Fig. 2.14, as the obtained flow is less than the sum of the lower flow bounds $\sum_{\tilde{l}_{ij}\neq\tilde{0}}\tilde{l}_{ij}$ in $\widetilde{G}^*$ ([27,33] < [40,49]).

Step 2. Define arc capacities of the fuzzy residual network $\widetilde{G}^{*\mu}$ according to the flow values going along the arcs in Fig. 2.14.

Construct fuzzy residual network, as shown in Fig. 2.15.

Step 3. Find the shortest path according to the number of arcs from s^* to t^* in the fuzzy residual network. Use the breadth-first-search and obtain the path $s^* \rightarrow x_4 \rightarrow t \rightarrow s \rightarrow x_2 \rightarrow t^*$.

Step 4. Pass $\tilde{\delta}^\mu = \min\left[\tilde{u}_{ij}^\mu\right]$, i.e. min ([13,16], [53,65], ∞, [48,60], [27,33]) = [13,16] flow units along the path $s^* \rightarrow x_4 \rightarrow t \rightarrow s \rightarrow x_2 \rightarrow t^*$

Step 5. Update the flow values in the graph $\widetilde{G}^*$.

The flow $\tilde{\xi}_{ij}^* = [27,33]$ turns to $\tilde{\xi}_{ij}^* = [27,33] + [13,16] = [40,49]$. Build the graph with the new flow value, as shown in Fig. 2.16.

Step 6.2. As the obtained flow equals the sum of the lower flow bounds $\sum_{\tilde{l}_{ij}\neq\tilde{0}}\tilde{l}_{ij}$ in $\widetilde{G}^*$ ([40,49] = [40,49]), the maximum flow in $\widetilde{G}^*$ and, therefore, the feasible flow in $\widetilde{G}$, defined by the flow passing along the artificial reverse arc (t, s) are obtained.. Thus, the feasible flow in $\widetilde{G}$ is [40,49] units. Graph $\widetilde{G}\left(\tilde{\xi}\right)$ with the feasible flow, defined according to the *rule 2.2*, is presented in Fig. 2.17.

Step 7. Determine arc capacities of the fuzzy residual network $\widetilde{G}^\mu\left(\tilde{\xi}\right)$ according to the rule *2.3* for the graph in Fig. 2.17 by the flow values, passing along the arcs of the graph.

Build fuzzy residual network, as shown in Fig. 2.18.

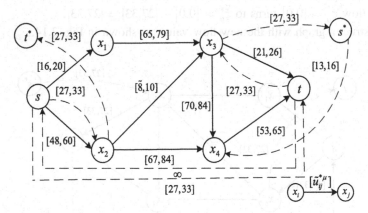

Fig. 2.15 Fuzzy residual network $\widetilde{G}^{*\mu}$ for the graph in Fig. 2.14

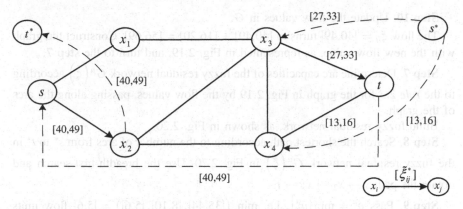

Fig. 2.16 Graph $\widetilde{G}^*$ with the new flow value of [40,49] units

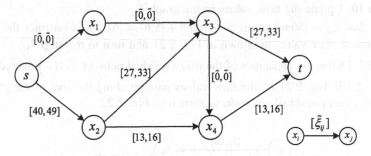

Fig. 2.17 Graph $\widetilde{G}(\widetilde{\xi})$ with the feasible flow of [40,49] units

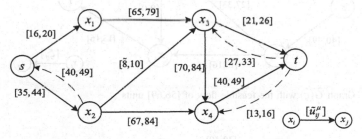

Fig. 2.18 Fuzzy residual network $\widetilde{G}^{\mu}(\widetilde{\xi})$ for the graph in Fig. 2.17

Step 8. Search the shortest path according to the number of arcs from s^* to t^* in the fuzzy residual network $\widetilde{G}^{\mu}\left(\widetilde{\xi}\right)$ in Fig. 2.18. Use the breadth-first-search and obtain the path $s \rightarrow x_1 \rightarrow x_3 \rightarrow t$.

Step 9. Pass $\widetilde{\delta}^{\mu} = \min\left[\widetilde{u}_{ij}^{\mu}\right]$, i.e. min ([16,20], [65,79], [21,26]) = [16,20] flow units along the path $s \rightarrow x_1 \rightarrow x_3 \rightarrow t$

Step 10. Update the flow values in $\widetilde{G}$.

The flow $\widetilde{\xi}_{ij} = [40,49]$ turns to $[40,49] + [16,20] = [56,69]$. Construct the graph with the new flow value, as represented in Fig. 2.19, and turn to the **step 7**.

Step 7. Determine arc capacities of the fuzzy residual network $\widetilde{G}^{\mu}\left(\widetilde{\xi}\right)$ according to the *rule 2.3* for the graph in Fig. 2.19 by the flow values, passing along the arcs of the graph.

Build fuzzy residual network, as shown in Fig. 2.20.

Step 8. Search the shortest path according to the number of arcs from s^* to t^* in the fuzzy residual network $\widetilde{G}^{\mu}\left(\widetilde{\xi}\right)$ in Fig. 2.20. Use the breadth-first-search and obtain the path $s \rightarrow x_2 \rightarrow x_3 \rightarrow t$.

Step 9. Pass $\widetilde{\delta}^{\mu} = \min\left[\widetilde{u}_{ij}^{\mu}\right]$, i.e. min $\left([35,44], [\widetilde{8},10], [\widetilde{5},\widetilde{6}]\right) = [\widetilde{5},\widetilde{6}]$ flow units along the path $s \rightarrow x_2 \rightarrow x_3 \rightarrow t$.

Step 10. Update the flow values in the graph $\widetilde{G}$.

The flow $\widetilde{\xi}_{ij} = [56,69]$ turns to $[56,69] + [\widetilde{5},\widetilde{6}] = [61,75]$. Construct the graph with the new flow value, as shown in Fig. 2.21 and turn to the **step 7**.

Step 7. Define arc capacities of the fuzzy residual network $\widetilde{G}^{\mu}\left(\widetilde{\xi}\right)$ according to the *rule 2.3* in Fig. 2.21 by the flow values passing along the arcs of the graph.

Build a fuzzy residual network, as shown in Fig. 2.22.

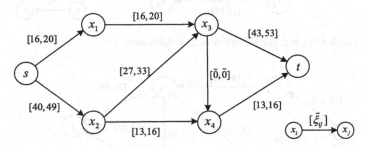

Fig. 2.19 Graph $\widetilde{G}(\widetilde{\xi})$ with the feasible flow of [56,69] units

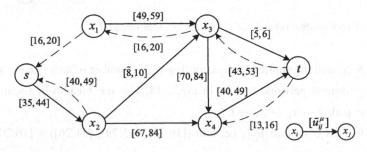

Fig. 2.20 Fuzzy residual network $\widetilde{G}^{\mu}(\xi)$ for the graph in Fig. 2.19

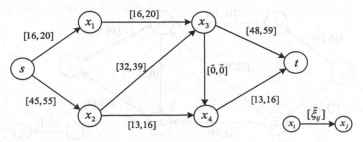

Fig. 2.21 Graph $\widetilde{G}(\tilde{\xi})$ with the feasible flow of [61,75] units

Step 8. Search the shortest path according to the number of arcs from s^* to t^* in the fuzzy residual network $\widetilde{G}^\mu\left(\tilde{\xi}\right)$ in Fig. 2.22. Use the breadth-first-search and obtain the path $s \rightarrow x_2 \rightarrow x_4 \rightarrow t$.

Step 9. Pass $\tilde{\delta}^\mu = \min\left[\tilde{u}_{ij}^\mu\right]$, i.e. min [30,38], [67,84], [40,49] = [30,38] flow units along the path $s \rightarrow x_2 \rightarrow x_4 \rightarrow t$.

Step 10. Update the flow values in the graph $\widetilde{G}\left(\tilde{\xi}\right)$.

Construct the graph with the new flow value, as shown in Fig. 2.23 and turn to the **step 7**.

Step 7. Define arc capacities of the fuzzy residual network $\widetilde{G}^\mu\left(\tilde{\xi}\right)$ according to the *rule 2.3* for the graph in Fig. 2.23 by the flow values, passing along the arcs of the graph.

Construct fuzzy residual network, as represented in Fig. 2.24.

Step 8. Search the shortest path according to the number of arcs from s^* to t^* in the fuzzy residual network $\widetilde{G}^\mu\left(\tilde{\xi}\right)$ in Fig. 2.24. Use the breadth-first-search and find that there is no such a path. Therefore, the maximum flow $\tilde{\xi}_{ij} + \tilde{\delta}^\mu \widetilde{P}^\mu = \tilde{v}$ of the value [91,113] flow units is obtained in the initial graph. Network with the maximum flow is represented in Fig. 2.23.

Define the borders of uncertainty of the resulting fuzzy interval [91,113] corresponding to the maximum flow in a graph $\widetilde{G}$. The obtained result follows the

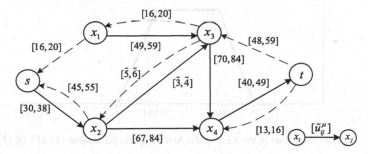

Fig. 2.22 Fuzzy residual network $\widetilde{G}^\mu(\xi)$ for the graph in Fig. 2.21

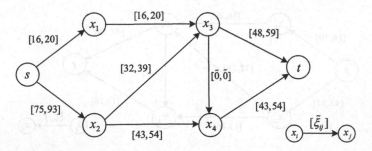

Fig. 2.23 Graph $\widetilde{G}(\tilde{\xi})$ with the feasible flow of [91,113] units

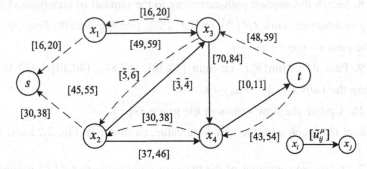

Fig. 2.24 Fuzzy residual network $\widetilde{G}^\mu(\tilde{\xi})$ for the graph in Fig. 2.23

basic value of fuzzy arc capacity: [74,89] with the left deviation $l_1^L = 16$ and the right deviation $l_1^R = 18$ in the form of the fuzzy trapezoidal number (74,89,16,18). Therefore, the deviation borders of the fuzzy interval [91,113] coincide with the deviation borders of the previous number.

Therefore, we can present the value of the maximum flow in the form of the trapezoidal fuzzy number (91,113,16,18), as shown in Fig. 2.25.

Define the borders of membership of the resulting fuzzy interval [91,113] corresponding to the maximum flow in a graph $\widetilde{G}$. The obtained membership follow the

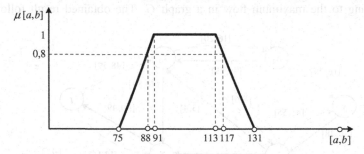

Fig. 2.25 The maximum flow in the form of the fuzzy trapezoidal number (91,113,16,18) of the flow units

The maximum flow with the degree of confidence equals 0,8 is in the interval [88,117], but anyway the maximum flow will be no less than 75 and no more than 131 units.

2.4 Minimum Cost Flow Finding in a Network with Fuzzy Arc Capacities and Transmission Costs

Suppose it is necessary to determine the minimum transportation cost of a certain quantity of cargo from a given point to the terminal one, given the restrictions on arc capacities of the road sections. Then we come to the problem of the minimum cost flow determining in the transportation network. Considering the fuzzy nature of arc capacities and transmission cost, we obtain the problem of determining the minimum cost flow in the transportation network in fuzzy conditions. Thus, the problem statement of the minimum cost flow finding in the transportation network with fuzzy arc capacities and costs can be represented as follows [4]:

$$\sum_{(x_i,x_j)\in \tilde{A}} \tilde{c}_{ij}\tilde{\xi}_{ij} \to \min, \tag{2.9}$$

$$\sum_{x_j\in \Gamma(x_i)} \tilde{\xi}_{ij} = \sum_{x_k\in \Gamma^{-1}(x_i)} \tilde{\xi}_{ki} = \begin{cases} \tilde{\rho}, & x_i = s, \\ -\tilde{\rho}, & x_i = t, \\ \tilde{0}, & x_i \neq s,t, \end{cases} \tag{2.10}$$

$$\tilde{\xi}_{ij} \leq \tilde{u}_{ij}, \quad \forall (x_i,x_j) \in \tilde{A}. \tag{2.11}$$

In the model (2.9)–(2.11) $\tilde{c}_{ij}$—transmission cost one the one flow unit along the arc (x_i,x_j); $\tilde{\rho}$—given flow value, which transmission cost should be minimized.

Basaker and Gowan's algorithms [12] of the sequential search of the shortest paths and M. Klein's algorithm of the negative weight cycles detection and removal are used for solving this problem in clear terms, as was described in the first chapter

Fig. 2.26 Initial network $\tilde{G}$

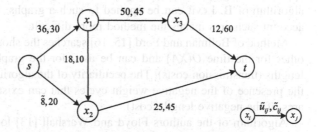

of the following monograph. Since the area of research is transportation networks, the original flow transmission costs are non-negative, then, the original graph will contain no cycles of the negative cost, therefore, it is inappropriate to apply M. Klein's at the first step of the minimal cost flow algorithm. It is also worth to notice that Floyd and Warshall's algorithm [13] operating routes from each node to any node is traditionally applied for detection of the negative cost cycles, which is not effective in this case because the it is sufficient to find the distance from the source to the sink. Therefore, P. Basaker, R. Gowan's algorithm and its modifications should be considered to find the minimum cost flow.

The search of the minimum cost path and maximum flow transmission along it, taking into account capacities of the arcs included in this path is carried out in the P. Basaker, R. Gowan's minimum cost flow algorithm. As the search of the minimum cost path is equivalent to finding the shortest path in the graph (in this case, the cost is equivalent to the length of the path), there is a question of choosing the optimal algorithm of the shortest path searching in the graph.

There are many algorithms for finding the shortest path, such as algorithms by Dijkstra [14], Bellman and Ford [15, 16], Floyd and Warshall [13], Levit [17], Johnson [18]. Traditionally, Dijkstra's algorithm [14] proposed in 1959 is used for finding the shortest paths (paths of the minimum cost), it is in searching for the shortest path from a given node to all the other vertices of the graph.

There are three sets that are supported during the algorithm: nodes distance to which has already been calculated (but perhaps not entirely); nodes, distance to which is calculated; nodes distance to which has not been not yet calculated. Time complexity of E. Dijkstra's algorithm depends on the data structure used to implement the queues with priority and representation of the input graph. In general, its running-time is $O(X^2)$ [19]. Algorithm of Levit [17] is also looking for the shortest path from a given node to all other vertices of the graph. Its running-time is exponential in the worst case, but in practice usually indicates time [19]. There are the same three sets that are supported during the algorithm. The nodes included in the first set are divided into two ordered sets—basic and priority queue. Each vertex is associated with a non-negative value of the length of the shortest of the currently known ways in it from the initial node. The disadvantage of the B. Levit's method is the necessity of the reprocessing of the nodes, while the advantage is the best running-time in graphs based on the real networks. But both methods are not applied to the graphs with the negative arcs lengths (transmission costs). In general, algorithm of B. Levit can be applied to such a graphs, but necessity to take into account such arcs makes the method more difficult.

Method of Bellman and Ford [15, 16] searches the shortest path from the node to other for the time $O(XA)$ and can be used for the graphs with the negative arcs lengths (transmission costs). The peculiarity of the algorithm is that it detects either the presence of the negative weight cycles that can exist if the original graph has arcs of the negative length (cost).

Algorithm of the authors Floyd and Warshall [13] looks for the shortest paths between all pairs of vertices in time $O(X^3)$. The method can be applied to graphs with negative path length (costs). It is used either to identify the cycles of the negative cost (for example, in the search algorithm of the minimum cost flow of M. Klein). This algorithm allows finding the shortest distance between the nodes without saving paths.

Algorithm of Johnson [18] looks for the shortest path between all pairs of nodes in time $O(X^2 \log X + XA)$ in general. It can be used for graphs with negative path lengths (costs). The algorithm is in the use of E. Dijkstra's algorithm for finding the shortest path previously changing weights of the edges, escaping from the negative lengths (costs). The new weights are introduced using the method of R. Bellman and L. Ford.

We look for the shortest path in the fuzzy residual network in the algorithm of the minimum cost flow finding. Thus, arcs of the negative costs appear as in the case of the passing the flow along the arc $(x_i, x_j) \in \widetilde{A}$ with the transmission cost $\tilde{c}_{ij}$ the reverse arc $\left(x_j^\mu, x_i^\mu\right) \in \widetilde{A}^\mu$ appears in the fuzzy residual network with the cost $-\tilde{c}_{ij}$. Therefore, we cannot apply an effective E. Dijkstra's algorithm for finding the minimum cost path. Therefore, to find the minimum cost flow it is appropriate to apply the algorithm proposed by R. Busacker and P. Gowen, where L. Ford's algorithm is applied at the stage of finding the shortest chain, which allows operating negative parameters of costs, or algorithm, which reduces the negative numbers to non-negative.

The method which allows turning from negative numbers to non-negative, is based on the introduction of the potentials for each node of the graph. More effective algorithms for searching the given minimum cost flow are algorithms of the authors Edmonds and Karp [20] and Tomizawa [21], transforming the negative cost of transportations via node potentials to non-negative. The main idea of the method is that at each step of the fuzzy residual network construction transmission costs are recalculated based on the values of the potentials assigned to nodes of the graph such a way that they are non-negative. In turn, the node potentials are defined by finding the lengths of the shortest paths from the source to the specific node, where the length of the path is its transmission cost.

Thus, let us consider basic definitions of the minimum cost flow finding in a network with fuzzy arc capacities and costs.

2.4.1   Potential Method for the Minimum Cost Flow Finding in a Network with Fuzzy Arc Capacities and Transmission

Let us consider basic concepts used in the method of the minimum cost flow finding of Tomizawa's [21] based on the algorithm of the potentials and modified costs introduction described in [22] and modified for the case of application of fuzzy numbers.

The key concept of the method is the concept of fuzzy modified (reduced) costs (Definition 2.4) which is a modification of the definition, taken from [22].

Definition 2.4 Let $\tilde{\pi}(x_i)$, $x_i \in X, i = 1, \ldots, n$—some specified node weights (node potentials). We define the so-called fuzzy "modified" or reduced costs $\tilde{c}_{ij}^{\pi}$ associated with the arcs of the residual network, such as:

$$\tilde{c}_{ij}^{\pi} = \tilde{c}_{ij}^{\mu} - \tilde{\pi}(x_i) + \tilde{\pi}(x_j). \tag{2.12}$$

In (2.12) $\tilde{c}_{ij}^{\mu}$ is the transmission cost (x_i, x_j) in the fuzzy residual network $\tilde{G}^{\mu}$, i.e.

$$\tilde{c}_{ij}^{\mu} = \begin{cases} \tilde{c}_{ij}, & \text{if } (x_i, x_j) \in \tilde{A}^{\mu}, (x_i, x_j) \in \tilde{A}, \\ -\tilde{c}_{ji}, & \text{if } (x_i, x_j) \in \tilde{A}^{\mu}, (x_i, x_j) \notin \tilde{A}. \end{cases}$$

For some cycle $\tilde{H}^{\mu}$ in the residual network:

$$\sum_{\left(x_i^{\mu}, x_i^{\mu}\right) \in \tilde{H}^{\mu}} \tilde{c}_{ij}^{\pi} = \sum_{\left(x_i^{\mu}, x_i^{\mu}\right) \in \tilde{H}^{\mu}} \tilde{c}_{ij}^{\mu}.$$

According to the condition from [22], let us consider the Definition 2.5.

Definition 2.5 The following equality is true for any path from s to t :

$$\sum_{\left(x_i^{\mu}, x_i^{\mu}\right) \in \tilde{P}^{\mu}} \tilde{c}_{ij}^{\pi} = \tilde{\pi}(t) - \tilde{\pi}(s) + \sum_{\left(x_i^{\mu}, x_i^{\mu}\right) \in \tilde{P}^{\mu}} \tilde{c}_{ij}^{\mu}. \tag{2.13}$$

According to condition (2.13) node potentials can't change the path of the minimum cost between any pair of nodes.

To prove the optimality of the flow, obtained after the introduction of the potentials, let us present Theorem 2.3, which is a modification of the theorem presented in [22].

Theorem 2.3 *The flow* $\tilde{\xi}^{*}$ *is optimal when and only when there is a vector potential, such as* $\tilde{c}_{ij}^{\pi} \geq \tilde{0}$ *for all* $\left(x_i^{\mu}, x_j^{\mu}\right) \in \tilde{A}^{\mu}$.

Proof Let us assume that there are such values $\tilde{\pi}$, that $\tilde{c}_{ij}^{\pi} \geq \tilde{0}$ for $\forall \left(x_i^{\mu}, x_j^{\mu}\right) \in \tilde{A}^{\mu}$. Let $\tilde{H}^{\mu}$ be a cycle in $\tilde{G}^{\mu}$, then:

$$\tilde{c}\left(\tilde{H}^{\mu}\right) = \sum_{\left(x_i^{\mu}, x_i^{\mu}\right) \in \tilde{H}^{\mu}} \tilde{c}_{ij}^{\mu} = \sum_{\left(x_i^{\mu}, x_i^{\mu}\right) \in \tilde{H}^{\mu}} \tilde{c}_{ij}^{\pi} \geq \tilde{0}.$$

Consequently, a negative cost cycle in $\widetilde{G}^{\mu}$ does not exist and, consequently, the flow is optimal. The theorem is proved.

Let us show that if there is no cycle of the negative cost in the residual network $\widetilde{G}^{\mu}$, then there is a flow vector with non-negative reduced costs.

Assume all nodes in the residual network $\widetilde{G}^{\mu}$ can be reachable from s. Let $\tilde{\gamma}^{\mu}(s, x_j)$ determines the length of the shortest path from s to x_j^{μ} in $\widetilde{G}^{\mu}$, assuming $\tilde{c}_{ij}^{\mu}$ are path lengths. If $\widetilde{G}^{\mu}$ has no cycle of the negative cost, then $\tilde{\gamma}^{\mu}(s, x_j) \leq \tilde{\gamma}^{\mu}(s, x_i) + \tilde{c}_{ij}^{\mu}$. Therefore, setting $\tilde{\pi}(x_j) = -\tilde{\gamma}^{\mu}(s, x_j)$ we obtain:

$$\tilde{c}_{ij}^{\pi} = \tilde{c}_{ij}^{\mu} - \tilde{\pi}(x_i) + \tilde{\pi}(x_j) = \tilde{c}_{ij}^{\mu} + \tilde{\gamma}^{\mu}(s, x_i) - \tilde{\gamma}^{\mu}(s, x_j) \geq \tilde{0}.$$

For flow $\tilde{\xi}$ and node $x_j \in X$ the node condition x_j is set as:

$$\tilde{e}(x_j) = \tilde{\rho}_j + \sum_{(x_i, x_j) \in \widetilde{A}} \tilde{\xi}_{ij} - \sum_{(x_j, x_i) \in \widetilde{A}} \tilde{\xi}_{ji}. \tag{2.14}$$

According to (2.14):

If $\tilde{e}(x_j) > \tilde{0}$, then the condition in the node is excess.

If $\tilde{e}(x_j) > \tilde{0}$, then the condition in the node is deficit.

If $\tilde{e}(x_j) = \tilde{0}$, then the condition in the node is balance.

$\tilde{\rho}_j$—the given flow value for the node x_j, called the balance of the node.

If $\tilde{\rho}_j > \tilde{0}$, then node x_j is called the source, $\tilde{\rho}_j$—supply of the node x_i.

If $\tilde{\rho}_j = \tilde{0}$, the node x_j is intermediate node.

If $\tilde{\rho}_j < \tilde{0}$, the node x_j is called a sink, $\tilde{\rho}_j$—the demand of the node x_i.

For further proof of the equality $\tilde{\pi}(x_j) = -\tilde{\gamma}^{\mu}(s, x_j)$ give Lemma 2.1, which is modification of the Lemma given in [22] for using in fuzzy conditions.

Lemma 2.1 *Let us assume that the flow $\tilde{\xi}$ and potentials, assigned to the nodes, satisfy the optimality criteria of reduced costs: $\tilde{c}_{ij}^{\pi} = \tilde{c}_{ij}^{\mu} - \tilde{\pi}(x_i) + \tilde{\pi}(x_j) \geq \tilde{0}$ for $\forall \left(x_i^{\mu}, x_j^{\mu} \right) \in \widetilde{A}^{\mu}$. For each node $x_j^{\mu} \in X^{\mu}$ let $\tilde{\gamma}^{\mu}(s, x_j)$ define the length of shortest path from s to x_j^{μ} in $\widetilde{G}^{\mu}$, assuming $\tilde{c}_{ij}^{\pi}$ as arc lengths.*

The flow $\tilde{\xi}$ satisfies the optimality criteria of the reduced costs relative to the potentials, i.e. the following condition is satisfied:

$$\tilde{\pi}'(x_j) = \tilde{\pi}(x_j) - \tilde{\gamma}^{\mu}(s, x_j) \quad \text{for } \forall\, x_j \in X. \tag{2.15}$$

Proof As $\tilde{\gamma}^\mu(s, x_j)$ are lengths of the shortest paths from s to x_j^μ in $\widetilde{G}^\mu$ in (2.15) for $\forall\left(x_i^\mu, x_j^\mu\right) \in \widetilde{A}$, then the following condition is satisfied:

$$\tilde{\gamma}^\mu(s, x_j) \le \tilde{\gamma}^\mu(s, x_i) + \tilde{c}_{ij}^\pi. \tag{2.16}$$

Therefore, based on (2.16):

$$\tilde{c}_{ij}^{\pi'} = \tilde{c}_{ij}^\mu - \tilde{\pi}'(x_i) + \tilde{\pi}'(x_j) = \tilde{c}_{ij}^\mu - (\tilde{\pi}(x_i) - \tilde{\gamma}^\mu(s, x_i)) + (\tilde{\pi}(x_j) - \tilde{\gamma}^\mu(s, x_j))$$
$$= \tilde{c}_{ij}^\pi + \tilde{\gamma}^\mu(s, x_i) - \tilde{\gamma}^\mu(s, x_j) \ge \tilde{0}.$$

Let us synthesize algorithm, which implements finding the minimum cost flow in the network with fuzzy arc capacities and costs, as well as the rule of the fuzzy residual network construction for finding the minimum cost flow, presented in [22] and modified for fuzzy conditions.

Rule 2.4 of construction of the fuzzy residual network for minimum cost flow finding in the network with fuzzy arc capacities and costs
Build fuzzy residual network $\widetilde{G}^\mu = (X^\mu, A^\mu)$, where $X^\mu = X$—the set of nodes of the residual network equals to the set of nodes X of the network $\widetilde{G}$, a $\widetilde{A}^\mu = \left\{ \left\langle \tilde{u}_{ij}^\mu / \left(x_i^\mu, x_j^\mu\right) \right\rangle \right\}$—fuzzy arcs set of the network $\widetilde{G}^\mu$, according to the following rules: for all arcs: if $\tilde{\xi}_{ij} < \tilde{u}_{ij}$, then include the corresponding arc in $\widetilde{G}^\mu$ with arc capacity $\tilde{u}_{ij}^\mu = \tilde{u}_{ij} - \tilde{\xi}_{ij}$, reduced cost $\tilde{c}_{ij}^\pi = \tilde{c}_{ij}^\mu - \tilde{\pi}_i + \tilde{\pi}_j$, where $\tilde{c}_{ij}^\pi = \tilde{c}_{ij}$. For all arcs, if $\tilde{\xi}_{ij} > \tilde{0}$, then include the corresponding arc in $\widetilde{G}^\mu$ with arc capacity $\tilde{u}_{ji}^\mu = \tilde{\xi}_{ij}$ and modified cost $\tilde{c}_{ji}^\pi = \tilde{c}_{ji}^\mu + \tilde{\pi}_i - \tilde{\pi}_j$, where $\tilde{c}_{ji}^\pi = -\tilde{c}_{ij}$.

Using the properties and lemmas presented below, we introduce algorithm for the minimum cost flow finding in the network with fuzzy arc capacities and transmission costs.

Algorithm of the minimum cost flow finding in the network with fuzzy arc capacities and transmission costs

Step 1. Introduce initial values: $\tilde{\xi}_{ij} = \tilde{0}, \tilde{\pi}(x_i) = \tilde{0}, \tilde{e}(x_i)\tilde{\rho}_i$ for $\forall x_i \in X$.

Step 2. Check the existence of the exceed in the node s.

2.1. If there is the exceed in the node s, i.e. $\tilde{e}(s) \ge \tilde{0}$, go to the **step 3**.

2.2. If there is no exceed in the node s, therefore, given flow value is found. To find its minimum cost turn to the given costs from the reduced ones, **exit**.

Step 3. Build fuzzy residual network $\widetilde{G}^\mu$ according to the *rule 2.4*. In the first step fuzzy residual network coincides with the original one due to the equality $\tilde{\xi}_{ij} = \tilde{0}$, and reduced costs coincide with initial ones, since $\tilde{\pi} = \tilde{0}$.

Step 4. Determine the minimum cost path $\widetilde{P}^\mu$ from s to t using E. Dijkstra's algorithm in the residual network, based on the reduced costs $\tilde{c}_{ij}^\pi$.

4.1. If the path exists, i.e. the permanent label is assigned to the node t, stop and go to the **step 5**.

4.2. If the path does not exist, i.e. vertex **t** is not reachable, then the task has no solution, **exit**.

Step 5. Pass $\tilde{\delta}^{\mu} = \min\{\tilde{u}(\tilde{P}^{\mu}), \tilde{e}(s), -\tilde{e}(t)\}$ flow units along the found path $\tilde{P}^{\mu}$, where $\tilde{u}(\tilde{P}^{\mu})$—arc capacity of the path $\tilde{P}^{\mu}$, defined by the minimum from the arc capacities of this path, i.e. $\tilde{u}(\tilde{P}^{\mu}) = \min\left[\tilde{u}_{ij}^{\mu}\right]$, $\left(x_i^{\mu}, x_j^{\mu}\right) \in \tilde{P}^{\mu}$.

Step 6. Define new values of the node potentials, as:

$$\tilde{\pi}(x_i) = \begin{cases} \tilde{\pi}(x_i) - \tilde{\gamma}^{\mu}(s, x_i), & \text{if the node } x_i^{\mu} \text{ has permanent label,} \\ \tilde{\pi}(x_i) - \tilde{\gamma}^{\mu}(s, t), & \text{in other case.} \end{cases} \qquad (2.17)$$

Step 7. Update flow values in the graph $\tilde{G}$: for arcs $\left(x_i^{\mu}, x_j^{\mu}\right) \notin \tilde{A}$, $\left(x_i^{\mu}, x_j^{\mu}\right) \in \tilde{A}^{\mu}$ in $\tilde{G}^{\mu}$ change the flow $\tilde{\xi}_{ji}$ along the corresponding arcs (x_j, x_i) of $\tilde{G}$ from $\tilde{\xi}_{ji}$ to $\tilde{\xi}_{ij} - \tilde{\delta}^{\mu}$. For the arcs $\left(x_i^{\mu}, x_j^{\mu}\right) \in \tilde{A}$, $\left(x_i^{\mu}, x_j^{\mu}\right) \in \tilde{A}^{\mu}$ in $\tilde{G}^{\mu}$ change the flow $\tilde{\xi}_{ij}$ along the corresponding arcs (x_i, x_j) of $\tilde{G}$ from $\tilde{\xi}_{ij}$ to $\tilde{\xi}_{ij} + \tilde{\delta}^{\mu}$, replace the flow value in the graph $\tilde{G}$: $\tilde{\xi}_{ij} \rightarrow \tilde{\xi}_{ij} + \tilde{\delta}^{\mu}\tilde{P}^{\mu}$ and turn to the **step 2**.

Since using the algorithm of E. Dijkstra we are interested in the way of the minimum cost from the source to the sink, there is no need to find the minimum cost path from the source to all vertices (as it is traditionally applied in the algorithm of E. Dijkstra). Consequently, the termination criterion of the minimum cost path algorithm of E. Dijkstra is not attributing the permanent marks to all vertices, and the attribution of the permanent mark to the sink node. Then updating the node potential, as it is presented in the **step 6** of the minimum cost flow algorithm will occur according to (2.17).

Let us show that this statement is true.

Thus, the new potential of the node x_i is replaced by the difference $\tilde{\pi}(x_i) - \tilde{\gamma}^{\mu}(s, x_i)$ in the shortest path from s to x_i^{μ}, if the node is assigned a permanent mark, which was shown in Lemma 2.1. If some vertices x_i have temporary labels, assign value $\tilde{\pi}(x_i) - \tilde{\gamma}^{\mu}(s, t)$ to the new values of the node potentials, i.e., the difference between the old value of potential and the shortest path length (path of the minimum cost) from s to **t**. This is true, because $\tilde{\gamma}^{\mu}(s, t)$ it is the greatest from the permanent marks, thus, these nodes, except t are not involved in the choice of the minimum cost path. That is, when the sink is reached, there is no need to consider nodes that haven't received permanent marks, so assign them the maximum of the obtained marks, i.e. $\tilde{\gamma}^{\mu}(s, t)$.

If it is necessary to find the maximum flow having a minimal cost, you can apply conventional algorithm of the minimum cost flow finding, but this algorithm requires preliminary assignment of supply and demand nodes. Since, the maximum required flow value is used as supply for the source and demand for the sink it is necessary to solve preliminary the maximum flow task in the fuzzy network for the supply and demand nodes assigning. Considering the computational complexity of J. Edmonds and R. Karp's algorithm for the maximum flow finding, it is advisable to consider the so-called algorithm of the subsequent shortest chain, described in [12] and the chain is found by Ford's algorithm [16].

Consider the numerical example that implements the algorithm of the minimum cost flow finding in the network with fuzzy arc capacities and costs.

Numerical example 3

Let the network is represented in the form of the fuzzy directed graph in Fig. 2.26. Values of arc capacities and transmission costs of the one flow unit are assigned to the arcs of the graph. It is necessary to define the minimum transportation cost of the fuzzy flow "near 18" units and represent the result in the form of the fuzzy triangular number, if there are basic values of arc capacities and transmission costs, represented in Figs. 2.27, 2.28, 2.29 and 2.30.

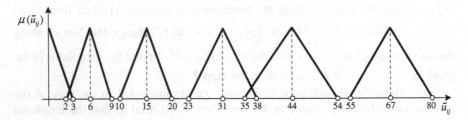

Fig. 2.27 Basic values of arc capacities

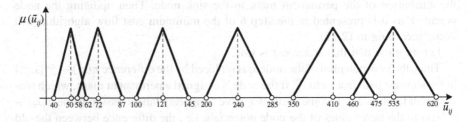

Fig. 2.28 Basic values of transmission costs

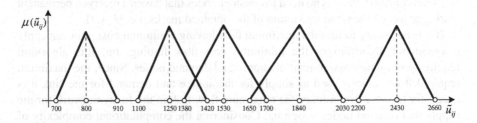

Fig. 2.29 Basic values of transmission costs

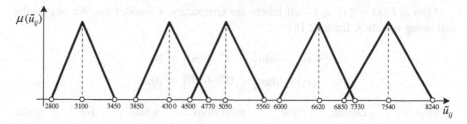

Fig. 2.30 Basic values of transmission costs

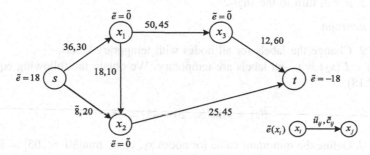

Fig. 2.31 Graph $\tilde{G}$ with given balances of the nodes

Step 1. Define initial values of flows and node potentials equal to $\tilde{0}$ i.e. $\tilde{\xi}_{ij} = \tilde{0}, \tilde{\pi}(x_i) = \tilde{0}$. Assign the corresponding value of the exceed to each node $\tilde{e}(x_j) = \tilde{p}_j + \sum_{(x_i,x_j)\in\tilde{A}} \tilde{\xi}_{ij} - \sum_{(x_j,x_i)\in\tilde{A}} \tilde{\xi}_{ji}$, as shown in Fig. 2.31.

Step 2.1. Define the condition of the node s: $\tilde{e}(s) = 18 + \tilde{0} - \tilde{0} > \tilde{0}$, i.e. the node has exceed, therefore, turn to the **step 3**.

Step 3. Fuzzy residual network $\tilde{G}^\mu = \left(X^\mu, \tilde{A}^\mu\right)$ coincides with the initial one, presented in Fig. 2.31, as the arc flows equal to $\tilde{0}$.

Step 4. Define the minimum cost path from s to t using E. Dijkstra's algorithm in the residual network, represented in Fig. 2.31.

Let $\tilde{l}(x_i) = \tilde{0}$—the label of the node x_i.

Step 1. Let $\tilde{l}(s) = \tilde{0}$ and consider this label as constant one. Let $\tilde{l}(x_i) = \infty$, $\forall x_i \neq s$ and consider these labels as temporary. Let $p = s$.

First iteration

Step 2. Change the labels for all nodes with temporary labels according to the following equality:

$$\tilde{l}(x_i) = \min\left[\tilde{l}(x_i), \tilde{l}(p) + \tilde{c}(p, x_i)\right]. \tag{2.18}$$

$\Gamma(p) = \Gamma(s) = \{x_1, x_2\}$—all labels are temporary. Consider x_1. We obtain the following equation from (2.18):

$$\tilde{l}(x_1) = \min\left[\infty, \tilde{0}^+ + 30\right] = 30,$$
$$\tilde{l}(x_2) = \min\left[\infty, \tilde{0}^+ + 20\right] = 20.$$

Step 3. Define the minimum value for nodes x_1, x_2, x_3, t : $\min[30, 20, \infty, \infty] = 20$, that corresponds to x_2.

Step 4. Consider the label of the node x_2 as constant $\tilde{l}(x_2) = 20$, $p = x_2$.

Step 5. $p \neq t$, turn to the *step 2*.

Second iteration

Step 2. Change the labels for all nodes with temporary labels.

$\Gamma(p) = \Gamma(x_2) = t$—all labels are temporary. We obtain the following equation from (2.18):

$$\tilde{l}(t) = \min\left[\infty, 20 + 45\right] = 65.$$

Step 3. Define the minimum value for nodes x_1, x_3, t : $\min[30, \infty, 65] = 30$, that corresponds to x_1.

Step 4. Consider the label of the node x_1 as constant, $\tilde{l}(x_1) = 30^+$, $p = x_1$.

Step 5. $p \neq t$, turn to the *step 2*.

Third iteration

Step 2. Change the labels for all nodes with temporary labels.

$\Gamma(p) = \Gamma(x_1) = \{x_2, x_3\}$—only x_3 has temporary label. We obtain the following equation from (2.18):

$$\tilde{l}(x_3) = \min[\infty, 30 + 45] = 75.$$

Step 3. Define the minimum value for nodes x_3, t : $\min[75, 65] = 65$, that corresponds to t.

Step 4. Consider the label of the node t as constant, $\tilde{l}(t) = 65^+$, $p = t$.

Step 5. $p = t$, therefore, we obtain the path of the minimum cost $\tilde{l}(t) = 65$. Find this path according to the following equation:

$$\tilde{l}(x_i') + \tilde{c}(x_i', x_i) = \tilde{l}(x_i). \tag{2.19}$$

In (2.19) x_i'—the node, proceeding to the node x_i in the path of the minimum cost from s to x_i.

Define the node, proceeding to the node t in the path of the minimum cost from s to t.

$$\tilde{l}(x_3) + \tilde{c}(x_3, t) \neq \tilde{l}(t), \quad \text{since } 75 + 60 \neq 65,$$
$$\tilde{l}(x_2) + \tilde{c}(x_2, t) = \tilde{l}(t), \quad \text{since } 20 + 45 = 65.$$

Obtain the path $x_2 \rightarrow t$. Define the node, proceeding to the node x_2 in the path of the minimum cost from s to t.

$$\tilde{l}(x_1) + \tilde{c}(x_1, x_2) \neq \tilde{l}(x_2), \quad \text{since } 30 + 10 \neq 20,$$
$$\tilde{l}(s) + \tilde{c}(s, x_2) = \tilde{l}(x_2), \quad \text{since } \tilde{0} + 20 = 20.$$

Therefore, the path $s \rightarrow x_2 \rightarrow t$ of the cost 65 units is obtained

Step 5. Pass $\tilde{\delta}^\mu = \min\{\tilde{u}(\widetilde{P}^\mu), \tilde{e}(s), -\tilde{e}(t)\}$, i.e. $\tilde{\delta}^\mu = \min\{8, 25, 18, 18\} = \tilde{8}$ flow units along the found path.

Step 6. Define new values of the node potentials:

$$\tilde{\pi}(s) = \tilde{0} - \tilde{0} = \tilde{0},$$
$$\tilde{\pi}(x_1) = \tilde{0} - 30 = -30,$$
$$\tilde{\pi}(x_2) = \tilde{0} - 20 = -20,$$
$$\tilde{\pi}(x_3) = \tilde{0} - 65 = -65,$$
$$\tilde{\pi}(t) = \tilde{0} - 65 = -65.$$

Step 7. Update the flow values in $\widetilde{G}$.

The flow $\tilde{\xi}_{ij} = \tilde{0}$ turns to $\tilde{8}$ units. Build the graph with the new flow value, as shown in Fig. 2.32 and turn to the **step 2**.

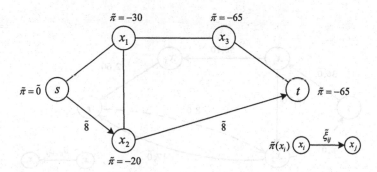

Fig. 2.32 Graph $\widetilde{G}$ with the flow of $\tilde{8}$ flow units and node potentials

Step 2. Check, if there is exceed in the node s. Since $\tilde{e}(s) = 18 + \tilde{0} - \tilde{8} > \tilde{0}$, i.e. the node has exceed, therefore, turn to the **step 3**.

Step 3. Calculate the new values of the reduced costs according to the paths lengths from s to t:

$$\tilde{c}^{\pi}_{sx_1} = \tilde{c}^{\mu}_{sx_1} - \tilde{\pi}_s + \tilde{\pi}_{x_1} = 30 - \tilde{0} - 30 - \tilde{0},$$

$$\tilde{c}^{\pi}_{x_2 s} = \tilde{c}^{\mu}_{x_2 s} - \tilde{\pi}_{x_2} + \tilde{\pi}_s = 20 + 20 + \tilde{0} = \tilde{0},$$

$$\tilde{c}^{\pi}_{x_1 x_2} = \tilde{c}^{\mu}_{x_1 x_2} - \tilde{\pi}_{x_1} + \tilde{\pi}_{x_2} = 10 + 30 - 20 = 20,$$

$$\tilde{c}^{\pi}_{x_1 x_3} = \tilde{c}^{\mu}_{x_1 x_3} - \tilde{\pi}_{x_1} + \tilde{\pi}_{x_3} = 45 + 30 - 65 = 10,$$

$$\tilde{c}^{\pi}_{x_2 t} = \tilde{c}^{\mu}_{x_2 t} - \tilde{\pi}_{x_2} + \tilde{\pi}_t = 45 + 20 - 65 = \tilde{0},$$

$$\tilde{c}^{\pi}_{t x_2} = \tilde{c}^{\mu}_{t x_2} - \tilde{\pi}_t + \tilde{\pi}_{x_2} = -45 + 65 - 20 = \tilde{0},$$

$$\tilde{c}^{\pi}_{x_3 t} = \tilde{c}^{\mu}_{x_3 t} - \tilde{\pi}_{x_3} + \tilde{\pi}_t = -60 + 65 - 65 = 60.$$

Build fuzzy residual network $\widetilde{G}^{\mu}$ according to the flow values passing along the arcs of the graph in Fig. 2.32 and calculated reduced costs, as shown in Fig. 2.33.

Step 4. Define the minimum cost path from s to t according to the E. Dijkstra's algorithm in the residual network, represented in Fig. 2.33: $s \rightarrow x_1 \rightarrow x_2 \rightarrow t$.

Step 5. Pass $\tilde{\delta}^{\mu} = \min\{\tilde{u}(\widetilde{P}^{\mu}), \tilde{e}(s), -\tilde{e}(t)\}$, i.e. $\tilde{\delta}^{\mu} = \min\{36, 18, 17, (18 + \tilde{0} - \tilde{8}), -(-18 + \tilde{8} - \tilde{0})\} = 10$ flow units along the found path.

Step 6. Define new values of the node potentials:

$$\tilde{\pi}(s) = \tilde{0} - \tilde{0} = \tilde{0},$$

$$\tilde{\pi}(x_1) = -30 - \tilde{0} = -30,$$

$$\tilde{\pi}(x_2) = -20 - 20 = -40,$$

$$\tilde{\pi}(x_3) = -65 - 10 = -75,$$

$$\tilde{\pi}(t) = -65 - 20 = -85.$$

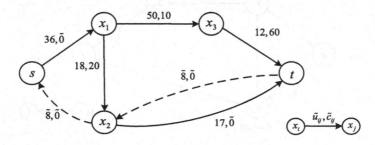

Fig. 2.33 Fuzzy residual network $\widetilde{G}^{\mu}$ for the graph in Fig. 2.32

Step 7. Update the flow values in $\widetilde{G}$.

The flow $\widetilde{\xi}_{ij} = \widetilde{8}$ turns to 18 units. Build the graph with the new flow value, as shown in Fig. 2.34 and turn to the **step 2**.

Step 2. Check, if there is exceed in the node s. Since $\tilde{e}(s) = 18 + \tilde{0} - 18 = \tilde{0}$, then the node s is balanced. Therefore, given flow value of the minimum cost is found.

Define the cost of the flow, which is equal to 18 flow units, as: $10 \cdot 30 + 8 \cdot 20 + 10 \cdot 10 + 18 \cdot 45 = 1370$ conventional units.

Let us define the uncertainty borders of the obtained fuzzy number 18, corresponded to the maximum flow in the graph $\widetilde{G}$. Found result is between two basic neighboring values of arc capacities: 15 with the left deviation $l_1^L = 5$, the right deviation—$l_1^R = 5$ and 31 with the left deviation $l_2^L = 8$, the right deviation—$l_2^R = 7$, presented in the form of fuzzy triangular numbers. According to (2.4) we obtain:

$$l^L = \frac{(a_2 - a')}{(a_2 - a_1)} \times l_1^L + \left(1 - \frac{(a_2 - a')}{(a_2 - a_1)}\right) \times l_2^L = \frac{(31 - 18)}{(31 - 15)} \times 5 + \left(1 - \frac{(31 - 18)}{(31 - 15)}\right) \times 8$$
$$= 5.5625 \approx 6,$$

$$l^R = \frac{(a_2 - a')}{(a_2 - a_1)} \times l_1^R + \left(1 - \frac{(a_2 - a')}{(a_2 - a_1)}\right) \times l_2^R = \frac{(31 - 18)}{(31 - 15)} \times 5 + \left(1 - \frac{(31 - 18)}{(31 - 15)}\right) \times 7$$
$$= 5.375 \approx 5.$$

Therefore, we can represent the value of the maximum flow in the form of the fuzzy triangular number (18,6,5), as shown in Fig. 2.35.

According to the Fig. 2.35 the flow (18,6,5) of the minimum cost with the degree of confidence equals to 0,7 should be in the interval [16,20], but anyway the flow of the minimum cost will be no less, than 12 units and no more than 23 units.

Let us represent the minimum transportation cost of (18,6,5) flow units equals to 1370 conventional units in the form of the fuzzy triangular number.

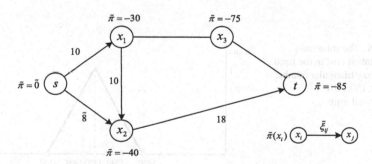

Fig. 2.34 Graph $\widetilde{G}$ with the flow of 18 units and node potentials

Fig. 2.35 Given flow in the form of the fuzzy triangular number of (18,6,5) units

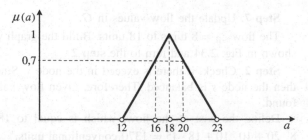

Let us define the uncertainty borders of the obtained fuzzy number 1370 conventional units. Found result is between two basic neighboring values of arc capacities: 1250 with the left deviation $l_1^L = 150$, right deviation—$l_1^R = 170$ and 1530 with the left deviation $l_1^L = 150$, right deviation—$l_1^R = 170$. According to (2.4) we obtain:

$$l^L = \frac{(a_2 - a')}{(a_2 - a_1)} \times l_1^L + \left(1 - \frac{(a_2 - a')}{(a_2 - a_1)}\right) \times l_2^L = \frac{(1530 - 1370)}{(1530 - 1250)} \times 150 + \left(1 - \frac{160}{280}\right) \times 150$$
$$= 150.2 \approx 150,$$

$$l^R = \frac{(a_2 - a')}{(a_2 - a_1)} \times l_1^R + \left(1 - \frac{(a_2 - a')}{(a_2 - a_1)}\right) \times l_2^R = \frac{(1530 - 1370)}{(1530 - 1250)} \times 170 + \left(1 - \frac{160}{280}\right) \times 170$$
$$= 170.1 \approx 170.$$

Therefore, we can represent the minimum flow transportation cost in the form of the fuzzy triangular number (1370,150,170), as shown in Fig. 2.36.

According to the Fig. 2.36 the minimum transportation cost of (18,6,5) flow units with the degree of confidence equals to 0,8 should be in the interval [1340,1404] conventional units, but anyway the minimum transportation cost of (18,6,5) units should be no less than 1220 and no more than 1540 conventional units.

Fig. 2.36 The minimum transportation cost in the form of the fuzzy triangular number of (1370,150,170) conventional units

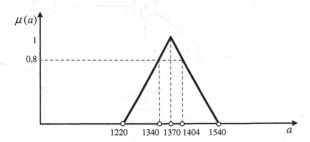

2.5 Minimum Cost Flow Finding in a Network with Fuzzy Nonzero Lower, Upper Flow Bounds and Transmission Costs

Fundamental problem, arising in the flow researches is the minimum cost flow finding in the network with nonzero lower flow bounds. This problem statement [23] is valid, when it is necessary to define the transportation routes of the minimum cost, taking into account transportation profitability set as the minimum flow value, that should be passed along the arcs.

In the real life it is impossible to consider this problem excluding changes in the environment and human activity, as such parameters of the network, as lower and upper flow bounds and transmission costs cannot be accurately measured. Therefore, these parameters should be represented in the fuzzy form. Thus, we obtain the problem statement of the minimum cost flow finding in the network with fuzzy nonzero lower, upper flow bounds and costs. This task will have a solution if the restriction on arc capacities will be satisfied for all arcs of the graph (2.22), and if the given flow value will not exceed the maximum flow in the graph. The problem statement can be represented as follows:

$$\sum_{(x_i, x_j) \in \tilde{A}} \tilde{c}_{ij} \tilde{\xi}_{ij} \to \min, \tag{2.20}$$

$$\sum_{x_j \in \Gamma(x_i)} \tilde{\xi}_{ij} = \sum_{x_k \in \Gamma^{-1}(x_i)} \tilde{\xi}_{ki} = \begin{cases} \tilde{\rho}, & x_i = s, \\ -\tilde{\rho}, & x_i = t, \\ \tilde{0}, & x_i \neq s, t, \end{cases} \tag{2.21}$$

$$\tilde{l}_{ij} \leq \tilde{\xi}_{ij} \leq \tilde{u}_{ij}, \quad \forall \, (x_i, x_j) \in \tilde{A}. \tag{2.22}$$

The flow task of the minimum cost flow finding in the network with fuzzy nonzero lower flow bounds, presented by the model (2.20)–(2.22) was considered in [10]. The methods of its solution in fuzzy conditions were not considered in the literature. Therefore, it is necessary to develop method, realized this task in fuzzy conditions.

To solve this problem it is necessary to introduce the rules of turning to the fuzzy graph without lower bounds flows (*the rule 2.5*) and search the maximum flow in it. The minimum cost flow is defined in the fuzzy residual network without lower flow bounds, constructed according to the Definition 2.6. After transition to the graph with the feasible flow, search given flow value of the minimum cost according to the construction of the fuzzy residual network with the feasible flow (*rule 2.6*). Present rules and definitions obtained by the synthesis of basic properties of the algorithm of the minimum cost flow finding with nonzero lower flow bounds, described in [10].

Rule 2.5 of turning to the corresponding fuzzy graph without lower flow bounds
from fuzzy graph with nonzero lower bounds flows for solving the minimum cost
flow problem with fuzzy nonzero lower flow bounds

Turn to the fuzzy graph $\widetilde{G}^* = (X^*, \widetilde{A}^*)$ without lower flow bounds from the given fuzzy graph $\widetilde{G} = (X, \widetilde{A})$ with nonzero lower bounds. Introduce artificial nodes s^* and t^*, the arc (t, s) with $\tilde{u}_{ts}^* = \infty$, $\tilde{l}_{ts}^* = \tilde{0}$, $\tilde{\xi}_{ij} + \tilde{\delta}^\mu \widetilde{P}^\mu$ in the new graph $\widetilde{G}$. For each node (x_i, x_j) in $\widetilde{G}$ with $\tilde{l}_{ij} \neq \tilde{0}$: (1) decrease $\tilde{u}_{ij}$ to $\tilde{u}_{ij}^* = \tilde{u}_{ij} - \tilde{l}_{ij}$, $\tilde{l}_{ij}$ to $\tilde{0}$, $\tilde{c}(\tilde{\xi}_{ij} + \tilde{\delta}^\mu \widetilde{P}^\mu)$. (2) Introduce arcs (s^*, x_j) and (x_i, t^*) with flow bounds equal to $\tilde{u}_{s^* x_j}^* = \tilde{u}_{x_i t^*}^* = \tilde{l}_{ij}$, $\tilde{l}_{s^* x_j}^* = \tilde{l}_{x_i t^*}^* = \tilde{0}$, costs $\tilde{c}_{s^* x_j}^* = \tilde{c}_{x_i t^*}^* = \tilde{0}$. Arcs without lower flow bounds are the same for $\widetilde{G}^*$: for any arc (x_i, x_j) with $\tilde{l}_{ij} = \tilde{0}$ is $\tilde{u}_{ij}^* = \tilde{u}_{ij}$, $\tilde{c}_{ij}^* = \tilde{c}_{ij}$.

Definition 2.6 *Fuzzy residual network for the graph $\widetilde{G}^* = (X^*, \widetilde{A}^*)$ without lower flow bounds for solving the minimum cost flow problem with fuzzy nonzero lower flow bounds—the network $\widetilde{G}^{*\mu} = (X^{*\mu}, \widetilde{A}^{*\mu})$, where $X^{*\mu} = X^*$—the set of the nodes of the fuzzy residual network with artificial nodes, $\widetilde{A}^{*\mu} = \left\{ <\tilde{u}_{ij}^{*\mu}/(x_i^{*\mu}, x_j^{*\mu}) > \right\}$—fuzzy set of the arcs of the network $\widetilde{G}^{*\mu}$, constructed according the following rules: for all arcs, if $\tilde{\xi}_{ij}^* < \tilde{u}_{ij}^*$, then include corresponding arc in $\widetilde{G}^{*\mu}$ with arc capacity $\tilde{u}_{ij}^{*\mu} = \tilde{u}_{ij}^* - \tilde{\xi}_{ij}^*$, modified cost $\tilde{c}_{ij}^{*\mu} = \tilde{c}_{ij}^*$. For all arcs, if $\tilde{\xi}_{ij}^* > \tilde{0}$, then include corresponding arc in $\widetilde{G}^{*\mu}$ with arc capacity $\tilde{u}_{ji}^{*\mu} = \tilde{\xi}_{ij}^*$, modified cost $\min[28, \tilde{8}] = \tilde{8}$.*

The rule 2.6 of the fuzzy residual network constructing with the feasible flow
vector for solving the minimum cost flow problem with fuzzy nonzero lower flow
bounds

For all arc, if $\tilde{\xi}_{s^* x_3}^* = \tilde{0}$, then include the corresponding arc in $\widetilde{G}^\mu(\tilde{\xi})$ with arc capacity $\tilde{u}_{ij}^\mu = \tilde{u}_{ij} - \tilde{\xi}_{ij}$, modified cost $\tilde{c}_{ij}^\mu = \tilde{c}_{ij}$. For all arc, if $\tilde{\xi}_{ij} > \tilde{l}_{ij}$, then include the corresponding arc in $\widetilde{G}^\mu(\tilde{\xi})$ with arc capacity $\tilde{u}_{ji}^\mu = \tilde{\xi}_{ij} - \tilde{l}_{ij}$, modified cost $\tilde{c}_{ji}^\mu = -\tilde{c}_{ij}$.

Based on the proposed rule represent algorithm for solving this task.

The minimum cost flow finding algorithm in the network with fuzzy non-zero lower, upper flow bounds and costs

Strep 1. Determine, if the task has a solution

1.1. If $\sum_{\tilde{l}_{ij} \neq \tilde{0}} \tilde{l}_{ij} \leq \tilde{\rho}$, turn to the **step 2**.

1.2. If $\sum_{\tilde{l}_{ij} \neq \tilde{0}} \tilde{l}_{ij} > \tilde{\rho}$, the task has no solution, **exit**.

Step 2. Let us define if the initial graph $\widetilde{G} = (X, \widetilde{A})$ has the feasible flow. Turn to the graph $\widetilde{G}^* = (X^*, \widetilde{A}^*)$ without lower flow bounds according to the *rule 2.5*.

Step 3. Build a fuzzy residual network starting with zero flows according to the Definition 2.6.

Step 4. Search the shortest path $\widetilde{P}^{*\mu}$ in terms of the number of arcs from the artificial source s^* to the artificial sink t^* in the constructed fuzzy residual network starting with zero flow values. The choice of the shortest path is according to the breadth-first search.

4.1. If the $\widetilde{P}^{*\mu}$ is found, go to the **step 5**.

4.2. The flow value $\tilde{\phi} < \sum_{\tilde{l}_{ij} \neq 0} \tilde{l}_{ij}$ is obtained, which is the maximum flow in $\widetilde{G}^*$, if the path is failed to find. It means that it is impossible to pass any unit of flow, but not all the artificial arcs are saturated. Therefore, initial graph $\widetilde{G}$ has no feasible flow and the task has no solution. Exit.

Step 5. Pass the minimum from the arc capacities $\tilde{\delta}^{*\mu} = \min[\tilde{u}(\widetilde{P}^{*\mu})]$, $\tilde{u}(\widetilde{P}^{*\mu}) = \min[\tilde{u}_{ij}^{*\mu}]$, $(x_i^{*\mu}, x_j^{*\mu}) \in \widetilde{P}^{*\mu}$ along the path $\widetilde{P}^{*\mu}$.

Step 6. Update the fuzzy flow values in the graph $\widetilde{G}^*$: change the fuzzy flow $\tilde{\xi}_{ji}^*$ along the corresponding arcs (x_j^*, x_i^*) from $\widetilde{G}^*$ by $\tilde{\xi}_{ji}^* - \tilde{\delta}^{*\mu}$ for arcs $(x_i^{*\mu}, x_j^{*\mu}) \notin \widetilde{A}^*$, $(x_i^{*\mu}, x_j^{*\mu}) \in \widetilde{A}^{*\mu}$ in $\widetilde{G}^{*\mu}$. Change the fuzzy flow $\tilde{\xi}_{ij}^*$ along the arcs (x_i^*, x_j^*) from $\widetilde{G}^*$ by $\tilde{\xi}_{ij}^* - \tilde{\delta}^{*\mu}$ for arcs $(x_i^{*\mu}, x_j^{*\mu}) \in \widetilde{A}^*$, $(x_i^{*\mu}, x_j^{*\mu}) \in \widetilde{A}^{*\mu}$ in $\widetilde{G}^{*\mu}$. Replace $\tilde{\xi}_{ij}^*$ by $\tilde{\xi}_{ij}^* + \tilde{\delta}^{*\mu} \widetilde{P}^{*\mu}$.

Step 7. Provide a comparison of the $\tilde{\xi}_{ij}^* + \tilde{\delta}^{*\mu} \widetilde{P}^{*\mu}$ и $\sum_{\tilde{l}_{ij} \neq \tilde{0}} \tilde{l}_{ij}$:

7.1. If the flow value $\tilde{\xi}_{ij}^* + \tilde{\delta}^{*\mu} \widetilde{P}^{*\mu}$ of the minimum cost $\tilde{c}(\tilde{\xi}_{ij}^* + \tilde{\delta}^{*\mu} \widetilde{P}^{*\mu})$ of the value $\tilde{\sigma}^*$ is less than $\sum_{\tilde{l}_{ij} \neq \tilde{0}} \tilde{l}_{ij}$ and less than given flow value $\tilde{\rho}$, i.e. $\tilde{\sigma}^* < \sum_{\tilde{l}_{ij} \neq \tilde{0}} \tilde{l}_{ij} \leq \tilde{\rho}$, i.e. not all artificial arcs become saturated, go to the **step 3**, i.e. to constructing of the new incremental graph with the flow passing along the arcs until $\tilde{\xi}_{ij}^* + \tilde{\delta}^{*\mu} \widetilde{P}^{*\mu}$ becomes equal to $\sum_{\tilde{l}_{ij} \neq \tilde{0}} \tilde{l}_{ij}$.

7.2. If the flow value $\tilde{\xi}_{ij}^* + \tilde{\delta}^{*\mu} \widetilde{P}^{*\mu}$ of the minimum cost $\tilde{c}\left(\tilde{\xi}_{ij}^* + \tilde{\delta}^{*\mu} \widetilde{P}^{*\mu}\right)$ of the value $\tilde{\sigma}^*$ is equal to $\sum_{\tilde{l}_{ij} \neq \tilde{0}} \tilde{l}_{ij}$ and no more than $\tilde{\rho}$, $\tilde{\sigma}^* = \sum_{\tilde{l}_{ij} \neq \tilde{0}} \tilde{l}_{ij} \leq \tilde{\rho}$, i.e. all artificial arcs become saturated, then the value $\tilde{\xi}_{ij}^* + \tilde{\delta}^{*\mu} \widetilde{P}^{*\mu}$ is required value of the maximum flow $\tilde{\sigma}^*$ of the minimum cost $\tilde{c}\left(\tilde{\xi}_{ij}^* + \tilde{\delta}^{*\mu} \widetilde{P}^{*\mu}\right)$. In this case, all arcs leaving the artificial source and entering artificial sink become saturated, and the flow along the artificial arc (t, s) in $\widetilde{G}^*$ determines the feasible flow in the initial graph $\widetilde{G}$ of the value $\tilde{\sigma} = \tilde{\xi}_{ts}^*$. Turn to the graph $\widetilde{G}$ from the graph $\widetilde{G}^*$ according to the *rule 2.2*. The network $\widetilde{G}(\tilde{\xi})$ with the feasible is obtained. Determine, if the flow is optimal.

7.2.1. If the flow in $\widetilde{G}(\tilde{\xi})$ is equal to $\tilde{\rho}$ of the cost $\sum_{(x_i, x_j) \in \widetilde{A}} \tilde{c}_{ij} \tilde{\xi}_{ij}$, we find the minimum cost flow, **the end**.

7.2.2. If the flow in $\widetilde{G}\left(\tilde{\xi}\right)$ is less than $\tilde{\rho}$ of the cost $\sum_{(x_i, x_j) \in \widetilde{A}} \tilde{c}_{ij} \tilde{\xi}_{ij}$, we obtain the network $\widetilde{G}\left(\tilde{\xi}\right)$ with the feasible flow. Turn to the **step 8**.

Step 8. Construct the residual network $\widetilde{G}^{\mu}\left(\tilde{\xi}\right)$ taking into account the feasible flow vector $\tilde{\xi} = \left(\tilde{\xi}_{ij}\right)$ in the graph $\widetilde{G}$ according to the *rule 2.6*.

Step 9. Define the shortest path $\widetilde{P}^\mu$ from the source to sink in the constructed residual network $\widetilde{G}^\mu\left(\widetilde{\xi}\right)$ according to L. Ford's algorithm.

9.1. Go to the **step 10** if the augmenting path $\widetilde{P}^\mu$ is found.

9.2. Given flow value exceeds the maximum flow in the graph $\widetilde{G}$ if the path is failed to find, i.e. $\widetilde{\rho} > \widetilde{v}$ and the task has no solution, **the end**.

Step 10. Pass $\widetilde{\delta}^\mu = \min\left[\widetilde{u}\left(\widetilde{P}^\mu\right)\right]$, $\widetilde{u}\left(\widetilde{P}^\mu\right) = \min\left[\widetilde{u}_{ij}^\mu\right]$, $\left(x_i^\mu, x_j^\mu\right) \in \widetilde{P}^\mu$ along the found path.

Step 11. Update the fuzzy flow values in the graph $\widetilde{G}$: replace the fuzzy flow $\widetilde{\xi}_{ji}$ along the corresponding arcs $\left(x_j, x_i\right)$ from $\widetilde{G}$ by $\widetilde{\xi}_{ji} - \widetilde{\delta}^\mu$ for arcs $\left(x_i^\mu, x_j^\mu \notin \widetilde{A}\right)$, $\left(x_i^\mu, x_j^\mu \in \widetilde{A}^\mu\right)$ in $\widetilde{G}^\mu\left(\widetilde{\xi}\right)$ and change the fuzzy flow $\widetilde{\xi}_{ij}$ along the arcs $\left(x_i, x_j\right)$ from $\widetilde{G}$ by $\widetilde{\xi}_{ij} + \widetilde{\delta}^\mu$ for arcs $\left(x_i^\mu, x_j^\mu \in \widetilde{A}\right)$, $\left(x_i^\mu, x_j^\mu\right) \in \widetilde{A}^\mu$ in $\widetilde{G}^\mu\left(\widetilde{\xi}\right)$ and replace the flow value in $\widetilde{G}$: $\widetilde{\xi}_{ij} \to \widetilde{\xi}_{ij} + \widetilde{\delta}^\mu \widetilde{P}^\mu$ and turn to the **step 12**.

Step 12. Compare the flow value $\widetilde{\xi}_{ij} + \widetilde{\delta}^\mu \widetilde{P}^\mu$ and $\widetilde{\rho}$:

12.1. If the flow value $\widetilde{\xi}_{ij} + \widetilde{\delta}^\mu \widetilde{P}^\mu$ of the minimum cost $\widetilde{c}\left(\widetilde{\xi}_{ij} + \widetilde{\delta}^\mu \widetilde{P}^\mu\right)$ is less than $\widetilde{\rho}$, then replace $\widetilde{\xi}_{ij}$ by $\widetilde{\xi}_{ij} + \widetilde{\delta}^\mu \widetilde{P}^\mu$ and turn to the **step 8**.

12.2. If the flow value $\widetilde{\xi}_{ij} + \widetilde{\delta}^\mu \widetilde{P}^\mu$ of the minimum cost $\widetilde{c}\left(\widetilde{\xi}_{ij} + \widetilde{\delta}^\mu \widetilde{P}^\mu\right)$ is equal to $\widetilde{\rho}$, therefore, the given flow value of the minimum cost is found, **the end**.

12.3. If the flow value $\widetilde{\xi}_{ij} + \widetilde{\delta}^\mu \widetilde{P}^\mu = \widetilde{h}$ of the minimum cost $\widetilde{c}\left(\widetilde{\xi}_{ij} + \widetilde{\delta}^\mu \widetilde{P}^\mu\right)$ is more than $\widetilde{\rho}$, then required flow is $\widetilde{\xi}_{ij} + \left(\widetilde{\delta}^\mu - \widetilde{h} + \widetilde{P}\right)\widetilde{P}^\mu$ of the minimum cost $\widetilde{c}\left(\widetilde{\xi}_{ij} + \left(\widetilde{\delta}^\mu - \widetilde{h} + \widetilde{P}\right)\widetilde{P}^\mu\right)$, **the end**.

Let us consider **step 9** of the algorithm in details. As the lengths of the shortest paths from s to any other node are found (they are the final label values), then paths can be obtained via recursive procedure using the following relation:

$$l(x_i) + \widetilde{c}(x_i, t) = l(t). \tag{2.23}$$

As the node x_i precedes the node t in the shortest path from s to t, then for any node x_i' the corresponding preceding arc x_i can be found as one of the remaining arcs, for which (2.23) is true. Due to the specific of flow tasks and rules of construction the incremental graphs (the reverse arc $\left(x_j^\mu, x_i^\mu\right)$ is introduced with arc capacity $\widetilde{u}_{ji}^\mu = \widetilde{\xi}_{ij}$, when the flow $\widetilde{\xi}_{ij} > \widetilde{0}$ is passing along the arc and artificial arcs of zero cost exist) finding the shortest path via recursive procedure, cycles can appear. If the node enters the cycle, the path is deleted and the alternative path of the minimum cost is selected (the alternative path cost and the path with cycles cost is equal).

Minimum cost maximum flow finding in the network with fuzzy nonzero lower, upper flow bounds and costs $\tilde{c}_{ij}^{\mu} \leq \tilde{0}$

Let it is necessary to find the minimum transmission cost of the maximum flow in the network taking into account restrictions on upper and lower flow bounds. This task isn't considered in the literature in fuzzy conditions, as lower, upper flow bounds and transmission costs can not be precisely defined. The range of factors, described in 1.2.1 influence parameters. Therefore, we come to the minimum cost maximum flow finding task with nonzero lower flow bounds in fuzzy conditions. The problem statement of this task is as follows:

$$\sum_{(x_i, x_j) \in \tilde{A}} \tilde{c}_{ij} \tilde{\xi}_{ij} \rightarrow \min, \tag{2.24}$$

$$\sum_{x_j \in \Gamma(x_i)} \tilde{\xi}_{ij} = \sum_{x_k \in \Gamma^{-1}(x_i)} \tilde{\xi}_{ki} = \begin{cases} \tilde{v}, & x_i = s, \\ -\tilde{v}, & x_i = t, \\ \tilde{0}, & x_i \neq s, t, \end{cases} \tag{2.25}$$

$$\tilde{l}_{ij} \leq \tilde{\xi}_{ij} \leq \tilde{u}_{ij}, \quad \forall \, (x_i, x_j) \in \tilde{A} \tag{2.26}$$

In the model (2.24)–(2.26) $\tilde{G}^{\mu}\left(\tilde{\xi}\right)$—the maximum fuzzy flow value, which transmission cost should be minimized.

Let us consider algorithm of the minimum cost maximum flow finding in the network with fuzzy nonzero upper, lower flow bounds and costs [24].

Step 1. Let us define if the initial graph $\tilde{G} = (X, \tilde{A})$ has the feasible flow. Turn to the graph $\tilde{G} = \left(X^*, \tilde{A}^*\right)$ without lower flow bounds according to the *rule 2.5*.

Step 2. Starting with zero flows, build a fuzzy residual network $\tilde{G}^{*\mu}$ starting with zero flows according to the Definition 2.6.

Step 3. Search the shortest path $\tilde{P}^{*\mu}$ in terms of the number of arcs from the artificial source s^* to the artificial sink t^* in the constructed fuzzy residual network starting with zero flow values. The choice of the shortest path is according to L. Ford's algorithm.

3.1. If the $\tilde{P}^{*\mu}$ is found, go to the **step 4**.

3.2. The flow value $\tilde{\phi}^* < \sum_{\tilde{l}_{ij} \neq \tilde{0}} \tilde{l}_{ij}$ is obtained, which is the maximum flow in $\tilde{G}^*$, if the path is failed to find. It means that it is impossible to pass any unit of flow, but not all the artificial arcs are saturated. Therefore, initial graph $\tilde{G}$ has no feasible flow and the task has no solution. Exit.

Step 4. Pass the minimum from the arc capacities $\tilde{\delta}^{*\mu} = \min[\tilde{u}(\tilde{P}^{*\mu})]$, $\tilde{u}(\tilde{P}^{*\mu}) = \min[\tilde{u}_{ij}^{*\mu}]$, $(x_i^{*\mu}, x_j^{*\mu}) \in \tilde{P}^{*\mu}$ along the path $\tilde{P}^{*\mu}$.

Step 5. Update the fuzzy flow values in the graph $\tilde{G}^*$: change the fuzzy flow $\tilde{\xi}_{ji}^*$ along the corresponding arcs (x_j^*, x_i^*) from $\tilde{G}^*$ by $\tilde{\xi}_{ji}^* - \tilde{\delta}^{*\mu}$ for arcs $(x_i^{*\mu}, x_j^{*\mu}) \notin \tilde{A}^*$,

$(x_i^{*\mu}, x_j^{*\mu}) \in \widetilde{A}^{*\mu}$ in $\widetilde{G}^{*\mu}$. Change the fuzzy flow $\widetilde{\xi}_{ij}^*$ along the arcs (x_i^*, x_j^*) from $\widetilde{G}^*$ by $\widetilde{\xi}_{ij}^* + \widetilde{\delta}^{*\mu}$ for arcs $(x_i^{*\mu}, x_j^{*\mu}) \in \widetilde{A}^*$, $(x_i^{*\mu}, x_j^{*\mu}) \in \widetilde{A}^{*\mu}$ in $\widetilde{G}^{*\mu}$. Replace $\widetilde{\xi}_{ij}^*$ by $\widetilde{\xi}_{ij}^* + \widetilde{\delta}^{*\mu} \widetilde{P}^{*\mu}$.

Step 6. Provide a comparison of the flow vector $\widetilde{\xi}_{ij}^* + \widetilde{\delta}^{*\mu} \widetilde{P}^{*\mu}$ and $\sum_{\tilde{l}_{ij} \neq \tilde{0}} \tilde{l}_{ij}$:

6.1. If the flow vector $\widetilde{\xi}_{ij}^* + \widetilde{\delta}^{*\mu} \widetilde{P}^{*\mu}$ of the minimum cost $\widetilde{c}(\widetilde{\xi}_{ij}^* + \widetilde{\delta}^{*\mu} \widetilde{P}^{*\mu})$ of the value $\widetilde{\sigma}^*$ is less than $\sum_{\tilde{l}_{ij} \neq \tilde{0}} \tilde{l}_{ij}$, i.e. not all artificial arcs become saturated, go to the **step 2**, i.e. to constructing of the new incremental graph with the flow passing along the arcs until $\widetilde{\xi}_{ij}^* + \widetilde{\delta}^{*\mu} \widetilde{P}^{*\mu}$ becomes equal to $\sum_{\tilde{l}_{ij} \neq \tilde{0}} \tilde{l}_{ij}$.

6.2. If the flow value $\widetilde{\xi}_{x_1 x_3}^* = \tilde{0} + \tilde{0} = \tilde{0}$ of the minimum cost $\widetilde{\xi}_{x_1 t^*}^* = \tilde{8} + \tilde{0} = \tilde{8}$ is equal to $\widetilde{\xi}_{x_2 x_3}^* = \tilde{0} + \tilde{0} = \tilde{0}$, i.e. all arcs from the artificial source to the artificial sink become saturated, then the value $\widetilde{\xi}_{x_2 t}^* = \tilde{0} + \tilde{0} = \tilde{0}$ is required value of the maximum flow $\widetilde{\xi}_{x_3 t}^* = \tilde{0} + \tilde{8} = \tilde{8}$ of the minimum cost $\widetilde{c}(\widetilde{\xi}_{ij}^* + \widetilde{\delta}^{*\mu} \widetilde{P}^{*\mu})$. In this case the flow passing along the artificial arc $\widetilde{\xi}_{ij}^* = \tilde{8}$ in $\widetilde{G}^*$ determines the feasible flow in the initial graph $\widetilde{G}^*$ of the value $\widetilde{\sigma} = \widetilde{\xi}_{ts}^*$. Turn to the graph $\widetilde{G}$ from the graph $\widetilde{G}^*$ according to the *rule 2.2*. The network is obtained. Go to the **step 7**.

Step 7. Construct the residual network $\widetilde{G}^\mu(\widetilde{\xi})$ taking into account the feasible flow vector $\widetilde{\xi} = (\widetilde{\xi}_{ij})$ in the graph $\widetilde{G}$ according to the *rule 2.6*.

Step 8. Define the shortest path $\widetilde{P}^\mu$ according to L. Ford's algorithm in the constructed residual network $\widetilde{G}^\mu(\widetilde{\xi})$.

8.1. Go to the **step 9** if the augmenting path $\widetilde{P}^\mu$ is found.

8.2. The maximum flow $\widetilde{\xi}_{ij} + \widetilde{\delta}^\mu \widetilde{P}^\mu = \widetilde{v}$ in $\widetilde{G}$ is found if the path is failed to find, then **stop**.

Step 9. Pass $\widetilde{\delta}^\mu = \min[\widetilde{u}(\widetilde{P}^\mu)]$, $\widetilde{u}(\widetilde{P}^\mu) = \min[\widetilde{u}_{ij}^\mu]$, $(x_i^\mu, x_j^\mu) \in \widetilde{P}^\mu$ along the found path.

Step 10. Update the fuzzy flow values in the graph $\widetilde{\xi}_{sx_2}^* < \widetilde{u}_{sx_2}^*$: replace the fuzzy flow $\widetilde{\xi}_{ji}$ along the corresponding arcs (x_j, x_i) from $\widetilde{G}$ by $\widetilde{\xi}_{ji} - \widetilde{\delta}^\mu$ for arcs $(x_i^\mu, x_j^\mu) \notin \widetilde{A}$, $(x_i^\mu, x_j^\mu) \in \widetilde{A}^\mu$ in $\widetilde{G}^\mu(\widetilde{\xi})$ and change the fuzzy flow $\widetilde{\xi}_{ij}$ along the arcs (x_i, x_j) from $\widetilde{G}$ by $\widetilde{\xi}_{ij} + \widetilde{\delta}^\mu$ for arcs $(x_i^\mu, x_j^\mu) \in \widetilde{A}$, $(x_i^\mu, x_j^\mu) \in \widetilde{A}^\mu$ in $\widetilde{G}^\mu(\widetilde{\xi})$ and replace the flow value in $\widetilde{G}$: $\widetilde{\xi}_{ij} \rightarrow \widetilde{\xi}_{ij} + \widetilde{\delta}^\mu \widetilde{P}^\mu$ and turn to the **step 7** starting from the new flow value along the arcs.

Give the proof of the main items of the algorithm. Introduce Theorem 2.4, that shows if the flow equals the sum of the lower flow bounds in $\widetilde{G}$ is found in $\widetilde{G}^*$, then original graph $\widetilde{G}$ has feasible flow.

Theorem 2.4 *If the maximum flow of the minimum cost in $\widetilde{G}^*$ is equal to the sum of the lower flow bounds $\widetilde{\sigma}^* = \sum_{\tilde{l}_{ij} \neq \tilde{0}} \tilde{l}_{ij}$, then the flow of the value $\widetilde{\sigma} = \widetilde{\xi}_{ts}^*$ exists in $\widetilde{G}$.*

Proof Proof that if the maximum flow in $\widetilde{G}^*$ is equal to the sum of the lower flow bounds $\tilde{\sigma}^* = \sum_{\tilde{l}_{ij} \neq \tilde{0}} \tilde{l}_{ij}$, then the flow of the value $\tilde{\sigma} = \xi^*_{ts}$ exists in $\widetilde{G}$ is presented in the Theorem 2.1. The required flow is the flow of the minimum cost according to the correctness of L. Ford's operations and equivalence of the minimum cost flow finding in $\widetilde{G}$ and $\widetilde{G}^*$. Thus, it is necessary to pass flow value equals the sum of the lower flow bounds along the paths of the minimum cost in $\tilde{8} < 36$, where the flow along the artificial arcs is used instead of lower flow bounds. If the feasible flow in $\widetilde{G}$ exists, then all arcs with nonzero lower flow bounds in $\widetilde{G}^{*\mu}$ should be saturated, i.e. the flow more, than lower flow bounds should pass along them. The costs of artificial arcs are equal to $\tilde{\xi}^*_{s^* x_1} < \tilde{u}^*_{s^* x_1}$, therefore, the choice of these arcs and passing the flow along them in $\tilde{u}^{*\mu}_{s^* x_1} = 28 - \tilde{8} = 20$ doesn't influence the minimum cost search in $\tilde{\xi}^*_{s^* x_1} > \tilde{0}$. Hence, we should pass the flow along the arcs with nonzero lower flow bounds, arcs from artificial source and arcs to artificial sink according to L. Ford's algorithm in $\tilde{u}^{*\mu}_{x_1 s^*} = \tilde{0} + \tilde{8} = \tilde{8}$ and obtain the minimum cost flow, that is equivalent to the search of the minimum cost flow in the initial graph and doesn't influence the choice of the minimum cost paths. Theorem is proved.

Let us consider numerical example, implementing the minimum cost flow finding algorithm with fuzzy nonzero lower, upper flow bounds and costs.

Numerical example 4
Let the network is represented in the form of the fuzzy directed graph in Fig. 2.37. Values of upper and lower flow bounds and transmission costs of one flow unit are assigned to arcs of the graph. It is necessary to determine the minimum transmission cost 36 flow units for the given graph with lower flow bounds and present results in the form of fuzzy triangular numbers or present, that this flow doesn't exist.

Step 1. Define, if the task has a solution.
Step 1.1. As $\sum_{\tilde{l}_{ij} \neq \tilde{0}} \tilde{l}_{ij} \leq \tilde{p}$, i.e. $36 = 36$, then turn to the **step 2**.
Step 2. Turn to the construction of the graph without lower flow bounds according to the *rule 2.5*, as shown in Fig. 2.38.

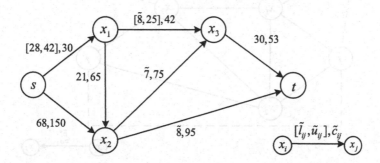

Fig. 2.37 Initial network $\widetilde{G}$

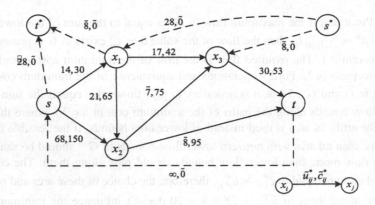

Fig. 2.38 Graph $\widetilde{G}^*$ without lower flow bounds

Step 3. Fuzzy residual network $\widetilde{G}^{*\mu}$ coincides with the graph $\widetilde{G}^*$ without lower flow bounds, presented in Fig. 2.38, as arc flows equal $\tilde{0}$.

Step 4. Find the shortest path according to the number of arcs from s^* to t^* in fuzzy residual network $\widetilde{G}^{*\mu}: s^* \rightarrow x_1 \rightarrow t^*$ of the cost $\tilde{0}$ conventional units.

Step 5. Pass $\tilde{\delta}^{*\mu} = \min\left[\tilde{u}_{ij}^{*\mu}\right]$, i.e. $\min[28,\tilde{8}] = \tilde{8}$ flow un its along the path $s^* \rightarrow x_1 \rightarrow t^*$

Step 6. Update the flow values in $\widetilde{G}^*$.

The flow $\tilde{\xi}_{ij}^* = \tilde{0}$ turns to $\tilde{\xi}_{ij}^* = \tilde{0} + \tilde{8} = \tilde{8}$.

Build a graph with the new flow value. As shown in Fig. 2.39.

Step 7.1. As the obtained flow is less than the sum of the lower flow bounds $\sum_{\tilde{l}_{ij} \neq \tilde{0}} \tilde{l}_{ij}$ in $\widetilde{G}^* (\tilde{8} < 36)$, turn to the **step 3**, i.e. to construction of the fuzzy residual network $\widetilde{G}^{*\mu}$ with the flow, presented in Fig. 2.39.

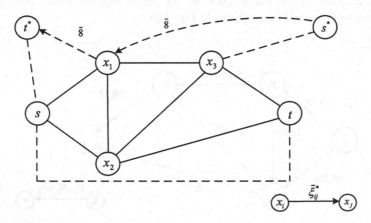

Fig. 2.39 Graph $\widetilde{G}^*$ with the new flow value of $\tilde{8}$ units

Step 3. Define arc capacities of the fuzzy residual network $\widetilde{G}^{*\mu}$ due to the flow values, passing along the arcs of the graph in Fig. 2.39.

Build fuzzy residual network, as shown in Fig. 2.40.

Step 4. Find the shortest path according to the number of arcs from s^* to t^* in $\widetilde{G}^{*\mu}$ due to the L. Ford's algorithm: $s^* \to x_3 \to t \to s \to t^*$ of the cost 53 conventional units.

Step 5. Pass $\tilde{\delta}^{*\mu} = \min\left[\tilde{u}_{ij}^{*\mu}\right]$, i.e. min $[\tilde{8}, 30, \infty, 28] = \tilde{8}$ flow units along the path $s^* \to x_3 \to t \to s \to t^*$.

Step 6. Update the flow values in $\widetilde{G}^*$.

The flow $\tilde{\xi}_{ij}^* = \tilde{8}$ turns to $\tilde{\xi}_{ij}^* = \tilde{8} + \tilde{8} = 16$.

Build graph with the new flow value, as shown in Fig. 2.41.

Step 7.1. Since the obtained flow is less than the sum of the lower flow bounds $\sum_{\tilde{l}_{ij} \neq \tilde{0}} \tilde{l}_{ij}$ in $\widetilde{G}^*$ ($16 < 36$), turn to the **step 3**.

Step 3. Determine arc capacities of the network $\widetilde{G}^{*\mu}$ due to the flow values, passing along the arcs of the graph in Fig. 2.41.

Build fuzzy residual network, as shown in Fig. 2.42.

Step 4. Find the shortest path according to the number of arcs from s^* to t^* in $\widetilde{G}^{*\mu}$ due to the L. Ford's algorithm: $s^* \to x_1 \to x_3 \to t \to s \to t^*$ of the cost of 95 units.

Step 5. Pass $\tilde{\delta}^{*\mu} = \min\left[\tilde{u}_{ij}^{*\mu}\right]$, i.e. min$[20, 17, 22, \infty, 20] = 17$ flow units along the path $s^* \to x_1 \to x_3 \to t \to s \to t^*$.

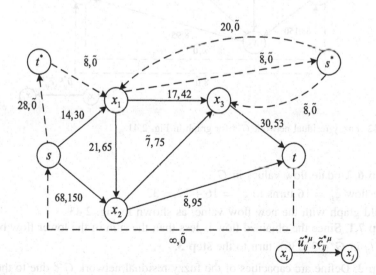

Fig. 2.40 Fuzzy residual network $\widetilde{G}^{*\mu}$ for graph in Fig. 2.39

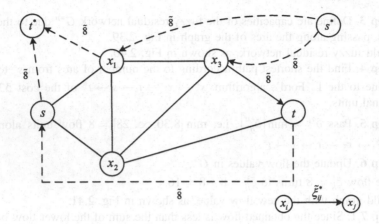

Fig. 2.41 Graph $\widetilde{G}^*$ with the new flow value of 16 units

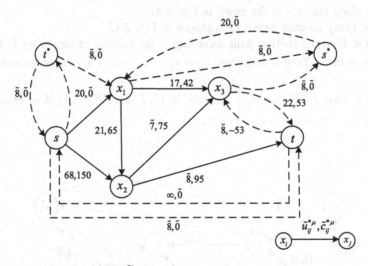

Fig. 2.42 Fuzzy residual network $\widetilde{G}^{*\mu}$ for graph in Fig. 2.41

Step 6. Update flow values in $\widetilde{G}^*$.

The flow $\tilde{\xi}_{ij}^* = 16$ turns to $\tilde{\xi}_{ij}^* = 16 + 17 = 33$.

Build graph with the new flow value, as shown in Fig. 2.43.

Step 7.1. Since the obtained flow is less than the sum of the lower flow bounds $\sum_{\tilde{l}_{ij} \neq \tilde{0}} \tilde{l}_{ij}$ in $\widetilde{G}^*$ (33 < 36), turn to the **step 3**.

Step 3. Define arc capacities of the fuzzy residual network $\widetilde{G}^{*\mu}$ due to the flow values, passing along the arcs of the graph in Fig. 2.43.

Build fuzzy residual network, as shown in Fig. 2.44.

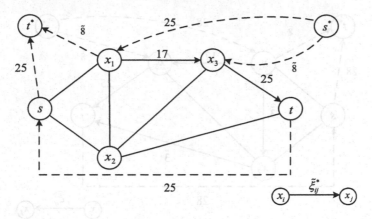

Fig. 2.43 Graph $\widetilde{G}^*$ with the new flow value of 33 units

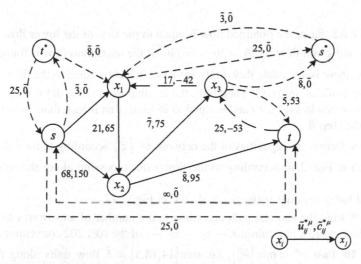

Fig. 2.44 Fuzzy residual network $\widetilde{G}^{*\mu}$ for the graph in Fig. 2.43

Step 4. Find the shortest path according to the number of arcs from s^* to t^* in $\widetilde{G}^{*\mu}$ due to the L. Ford's algorithm: $s^* \to x_1 \to x_2 \to t \to s \to t^*$ of the cost и 160 conventional units.

Step 5. Pass $\tilde{\delta}^{*\mu} = \min\left[\tilde{u}_{ij}^{*\mu}\right]$, i.e. min $[3, 21, \tilde{8}, \infty, \tilde{3}] = \tilde{3}$ flow units along the path $s^* \to x_1 \to x_2 \to t \to s \to t^*$.

Step 6. Update flow values in $\widetilde{G}^*$.

The flow $\tilde{\xi}_{ij}^* = 33$ turns to $\tilde{\xi}_{ij}^* = 33 + \tilde{3} = 36$. Build graph with the new flow value, as shown in Fig. 2.45.

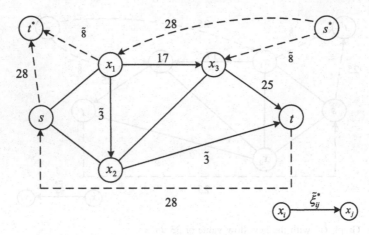

Fig. 2.45 Graph $\widetilde{G}^*$ with the new flow value of 36 units

Step 7.2.2. Since the obtained flow is equal to the sum of the lower flow bounds $\sum_{\tilde{l}_{ij} \neq \tilde{0}} \tilde{l}_{ij}$ and given flow $\tilde{\rho}$ ($36 = 36 = 36$) in $\widetilde{G}$, the maximum flow is found in $\widetilde{G}^*$, therefore, there is a feasible flow in the initial $\widetilde{G}$, which is equal to the flow, passing along the artificial arc (t, s). Therefore, the feasible flow in $\widetilde{G}$, defined by the *rule 2.2*, is presented in Fig. 2.46 and is equal to 28 units, that is less, than 36 units, thus, turn to the **step 8**.

Step 8. Define arc capacities of the network $\widetilde{G}^{\mu}\left(\tilde{\xi}\right)$ according to the *rule 2.6* for the graph in Fig. 2.46 according to the flow values, passing along the arcs of the graph.

Build fuzzy residual network, as shown in Fig. 2.47.

Step 9. Find the shortest path according to the number of arcs from s to t in $\widetilde{G}^{\mu}$ due to the L. Ford's algorithm: $s \rightarrow x_1 \rightarrow x_2 \rightarrow t$ of the cost 202 conventional units.

Step 10. Pass $\tilde{\delta}^{\mu} = \min\left[\tilde{u}_{ij}^{\mu}\right]$, i.e. $\min\left[14, 18, \tilde{5}\right] = \tilde{5}$ flow units along the path $s \rightarrow x_1 \rightarrow x_2 \rightarrow t$.

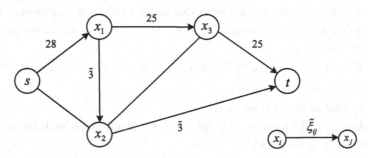

Fig. 2.46 Graph $\widetilde{G}(\tilde{\xi})$ with the feasible flow of units 28

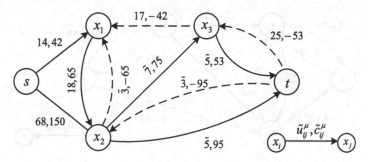

Fig. 2.47 Fuzzy residual network $\widetilde{G}^{\mu}(\widetilde{\xi})$ for the graph in Fig. 2.46

Step 11. Update the flow values in $\widetilde{G}\left(\widetilde{\xi}\right)$.

Step 12.1. Compare the flow value $\widetilde{\xi}_{ij} + \widetilde{\delta}^{\mu}\widetilde{P}^{\mu}$ and $\widetilde{\rho}$. As $33 < 36$ change $\widetilde{\xi}_{ij}$ by $\widetilde{\xi}_{ij} + \widetilde{\delta}^{\mu}\widetilde{P}^{\mu}$, as shown in Fig. 2.48.

The flow $\widetilde{\xi}_{ij} = 28$ turns to $\widetilde{\xi}_{ij} = 28 + \widetilde{5} = 33$, go to the **step 9**.

Step 8. Define arc capacities of the fuzzy residual network $\widetilde{G}^{\mu}\left(\widetilde{\xi}\right)$ due to the *rule 2.7* for the graph in Fig. 2.48 according to the flow values, passing along the arcs of the graph.

Build fuzzy residual network, as shown in Fig. 2.49.

Step 9. Find the shortest path according to the number of arcs from s to t in $\widetilde{G}^{\mu}$ due to the L. Ford's algorithm: $s \rightarrow x_1 \rightarrow x_2 \rightarrow x_3 \rightarrow t$ of the cost 223 conventional units.

Step 10. Pass $\widetilde{\delta}^{\mu} = \min\left[\widetilde{u}_{ij}^{\mu}\right]$, i.e. $\min\left[\widetilde{9}, 13, \widetilde{7}, \widetilde{5} = \widetilde{5}\right]$ flow units along the path $s \rightarrow x_1 \rightarrow x_2 \rightarrow x_3 \rightarrow t$.

Step 11. Update flow values in $\widetilde{G}\left(\widetilde{\xi}\right)$.

The flow $\widetilde{\xi}_{ij} = 33$ turns to $\widetilde{\xi}_{ij} = 33 + \widetilde{5} = 38$. Build a graph with the new flow value, as shown in Fig. 2.50 and turn to the **step 12**.

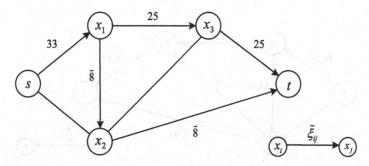

Fig. 2.48 Graph $\widetilde{G}(\widetilde{\xi})$ with the feasible flow of 33 units

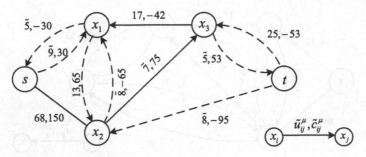

Fig. 2.49 Fuzzy residual network $\widetilde{G}^{\mu}(\widetilde{\xi})$ for the graph in Fig. 2.48

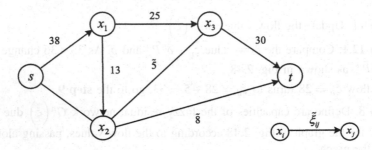

Fig. 2.50 Graph $\widetilde{G}(\widetilde{\xi})$ with the feasible flow of 38 units

Step 12.3. Compare flow value $\widetilde{\xi}_{ij} + \widetilde{\delta}^{\mu}\widetilde{P}^{\mu}$ and $\widetilde{\rho}$. Since $38 > 36$, then required flow value will be $33 + (\widetilde{5} - 38 + 36) \times \widetilde{P}^{\mu}$, i.e. $33 + \widetilde{3} \times \widetilde{P}^{\mu}$, where $\widetilde{P}^{\mu} = s \rightarrow x_1 \rightarrow x_2 \rightarrow x_3 \rightarrow t$, that presented in Fig. 2.51.

Minimum transmission cost of 36 flow units is $36 \cdot 30 + 11 \cdot 65 + 25 \cdot 42 + \widetilde{3} \cdot 75 + \widetilde{8} \cdot 95 + 28 \cdot 53 = 5314$ conventional units.

Let us define deviation borders of obtained fuzzy number of 36 units, that corresponds to the given flow value in $\widetilde{G}$. The found result is between two basic neighboring values of fuzzy arc capacities: 31 with the left deviation $l_1^L = 8$, right

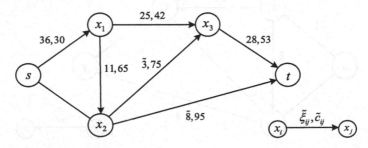

Fig. 2.51 Graph $\widetilde{G}(\widetilde{\xi})$ with the flow 36 units

deviation—$l_1^R = 7$ and 44 with the left deviation $l_2^L = 9$, right deviation—$l_2^R = 10$. According to (2.4) it is obtained:

$$l^L = \frac{(a_2 - a')}{(a_2 - a_1)} \times l_1^L + \left(1 - \frac{(a_2 - a')}{(a_2 - a_1)}\right) \times l_2^L = \frac{(44 - 36)}{(44 - 31)} \times 8 + (1 - \frac{(44 - 36)}{(44 - 31)}) \times 9$$
$$= 8.34 \approx 8,$$
$$l^R = \frac{(a_2 - a')}{(a_2 - a_1)} \times l_1^R + \left(1 - \frac{(a_2 - a')}{(a_2 - a_1)}\right) \times l_2^R = \frac{(44 - 36)}{(44 - 31)} \times 7 + (1 - \frac{(44 - 36)}{(44 - 31)}) \times 10$$
$$= 8.11 \approx 8.$$

Therefore, now we can represent the value of the given flow in the form of the fuzzy triangular number (36,8,8), as shown in Fig. 2.52.

Due to the Fig. 2.52 the minimum cost flow with the degree of confidence equals 0,6 will be within the interval [33,39] , but anyway it will no less than 28 units and no more than 44 units.

Let us represent the minimum transmission of (36,8,8) flow units equals 5314 conventional units in the form of the fuzzy triangular number.

Define uncertainty borders of the given fuzzy flow of 5314 conventional units. The obtained result is between two basic neighboring values of fuzzy transmission costs: 5050 with the left deviation $l_1^L = 550$, right deviation—$l_1^R = 510$ and 6620 with the left deviation $l_2^L = 620$, right deviation—$l_2^R = 710$. According to (2.4) it is obtained:

$$l^L = \frac{(a_2 - a')}{(a_2 - a_1)} \times l_1^L + \left(1 - \frac{(a_2 - a')}{(a_2 - a_1)}\right) \times l_2^L = \frac{(6620 - 5314)}{(6620 - 5050)} \times 550 + \left(1 - \frac{1306}{1570}\right) \times 620$$
$$= 562.9 \approx 563,$$
$$l^R = \frac{(a_2 - a')}{(a_2 - a_1)} \times l_1^R + \left(1 - \frac{(a_2 - a')}{(a_2 - a_1)}\right) \times l_2^R = \frac{(6620 - 5314)}{(6620 - 5050)} \times 510 + \left(1 - \frac{1306}{1570}\right) \times 710$$
$$= 544.9 \approx 545.$$

Therefore, it is to represent the minimum transportation cost in the form of the fuzzy triangular number (5314,563,545), as shown in Fig. 2.53.

Fig. 2.52 Given flow value in the form of the fuzzy triangular number of (36,8,8) units

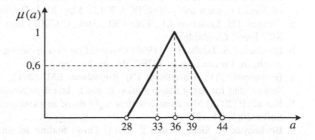

Fig. 2.53 Minimum
transmission cost in the form
of the fuzzy triangular number
of (5314,563,545)
conventional units

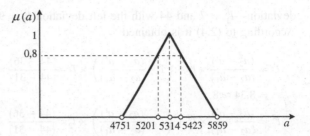

Due to the Fig. 2.53 the minimum transmission cost of (36,8,8) flow units with
the degree of confidence 0,8 will be within the interval of [5201,5423] conventional
units, but anyway the minimum transmission cost of (36,8,8) flow units will be no
less, than 4751 and no more, than 5859 conventional units.

2.6 Summary

This chapter deals with main concepts of fuzzy logic. The basic flow tasks in fuzzy
networks are described. The factors leading to the problem statements in fuzzy
conditions are given. The conclusion is in necessity of flow tasks considering in
fuzzy conditions.

References

1. Bozhenyuk A, Rozenberg I, Starostina T (2006) Analiz i issledovaniye potokov i zhivuchesti v
   transportnykx setyakh pri nechetkikh dannykh. Nauchnyy Mir, Moskva
2. Bershtein L, Bozhenuk A (2008) Fuzzy graphs and fuzzy hypergraphs. In: Dopico J, de la
   Calle J, Sierra A (eds) Encyclopedia of artificial intelligence information SCI. Hershey, New
   York, pp 704–709
3. Christofides N (1975) Graph theory: an algorithmic approach. Academic Press, New York
4. Bozhenyuk A, Gerasimenko E, Rozenberg I (2012) The methods of maximum flow and
   minimum cost flow finding in fuzzy network. In: Proceedings of the concept discovery in
   unstructured data (CDUD 2012): workshop co-located with the 10th international conference
   on formal concept analysis (ICFCA 2012), May 2012. Belgium, Leuven, pp 1–12
5. Cormen TH, Leiserson CE, Rivest RL, Stein C (2001) Introduction to algorithms, 3rd edn.
   MIT Press, Cambridge
6. Bershtein LS, Dzuba TA, (1998) Construction of a spanning subgraph in the fuzzy bipartite
   graph. In: Proceedings of EUFIT'98, Aachen, pp 47–51
7. Bozhenyuk AV, Rozenberg IN, Rogushina EM (2011) Approach of maximum flow
   determining for fuzzy transportation network. Izvestiya SFedU. Eng Sci 5(118):83–88
8. Kovács P (2015) Minimum-cost flow algorithms: an experimental evaluation. Optim Methods
   Softw 30(1):94–127
9. Bozhenyuk A, Gerasimenko E (2014) Flows finding in networks in fuzzy conditions. In:
   Kahraman C, Öztaysi B (eds) Supply chain management under fuzzines, studies in fuzziness

and soft computing. Springer, Berlin, vol 313, Part III, pp 269–291. doi:10.1007/978-3-642-53939-8_12

10. Murty KG (1992) Network programming. Prentice Hall, Upper Saddle River
11. Gerasimenko EM (2012) Minimum cost flow finding in the network by the method of expectation ranking of fuzzy cost functions. Izvestiya SFedU. Eng Sci 4(129):247–251
12. Busacker RG, Gowen P (1961) A procedure for determining a family of minimum-cost network flow patterns. In: Technical report 15, Operations Research Office, John Hopkins University
13. Floyd RW (1962) Algorithm 97: shortest path. Commun ACM 5(6):345
14. Dijkstra EW (1959) A note on two problems in connexion with graphs. Numer Math 1:269–271
15. Bellman R (1958) On a routing problem. Q Appl Math 16(1):87–90
16. Ford LR Jr (1956) Network flow theory. In: Paper P-923. RAND Corporation, Santa Monica, California
17. Levit BY (1971) Algoritmi poiska kratchajshikh putej na grahe. Trudi institute gidrodinamiki SO AN SSSR, Sb. "Modelirovaniye proczessov upravleniya, Novosibirsk, vip. 4, 1117–1148
18. Johnson DB (1977) Efficient algorithms for shortest paths in sparse networks. J ACM 24(1):1–13
19. Levitin A (2006) Introduction to the design and analysis of algorithms, 2nd edn. Addison-Wesley, Boston
20. Edmonds J, Karp RM (1972) Theoretical improvements in algorithmic efficiency for network flow problems. J Assoc Comput Mach 19(2):248–264
21. Tomizawa N (1971) On some techniques useful for solution of transportation network problems. Networks 1:173–194
22. Pouly A (2010) Minimum cost flows, 23 Nov 2010, http://perso.ens-lyon.fr/eric.thierry/Graphes2010/amaury-pouly.pdf
23. Bozhenyuk A, Gerasimenko E, Rozenberg I (2012) The task of minimum cost flow finding in transportation networks in fuzzy conditions. In: Proceedings of the 10th international FLINS conference on uncertainty modeling in knowledge engineering and decision making word scientific, Istanbul, Turkey, 26–29 Aug 2012, pp 354–359
24. Bozhenyuk AV, Gerasimenko EM, Rozenberg IN (2012) Algorithm of minimum cost fuzzy flow finding in a network with fuzzy arc capacities and transmission costs 2012. Izvestiya SFedU. Eng Sci 5(130):118–122

Chapter 3
Flow Tasks Solving in Dynamic Networks with Fuzzy Lower, Upper Flow Bounds and Transmission Costs

Tasks, considered in the first chapter of the present book, assume instant flow transition along the arcs of the graph. The present paper deals with dynamic networks, i.e. such a networks, in which flow spends certain time passing along the arcs of the graph. These problems arise in planning of transportation, economic planning, construction and evacuation of buildings. Methods of solving flow problems in dynamic transportation networks, in particular, the problem of the minimum cost flow dynamic in a dynamic network are used to solve the problem of air pollution by machines, connecting arc capacity with the maximum permissible level of air pollution, which in turn depends on the number of machines which can skip a particular track section [1]. These tasks should be considered in terms of the complex nature of the environmental factors that contribute uncertainty to the problem statements. The urgency of these tasks and similar tasks, taking into account the nonzero lower flow bounds, reflecting the economic feasibility of transportation, is that the algorithms of their solution allow us to find the maximum amount of traffic between selected points on the road map and choose the best routes for the transportation in terms of the limited transportation time and time-varying parameters of the network in term of uncertainty.

3.1 Definition of the Fuzzy Dynamic Network

As was mentioned in the first chapter, dynamic network is a network [2] $G = (X, A)$, where $X = \{x_1, x_2 \ldots, x_n\}$—the set of nodes, $A = \{(x_i, x_j)\}$, $i, j \in I = \overline{1, n}$—the set of arcs. Each arc of the dynamic graph (x_i, x_j) is set two parameters: transit time τ_{ij} and arc capacity u_{ij}. The time horizon $T = 0, 1, \ldots, p$ determining that all flow units sent from the source must arrive at the sink within the time p is given [2]. Let τ_{ij} be a positive number. Let $\Gamma(x_i)$ is the set of nodes, arcs from x_i go to, $\Gamma^{-1}(x_i)$ is the set of nodes, arcs from x_i go from. Thus, not more

© Springer International Publishing Switzerland 2017
A.V. Bozhenyuk et al., *Flows in Networks Under Fuzzy Conditions*,
Studies in Fuzziness and Soft Computing 346, DOI 10.1007/978-3-319-41618-2_3

than u_{ij} units of flow can be sent along the arc (x_i, x_j) at each time period in dynamic networks. Let x_j be the terminal node and x_i is the initial node of the arc (x_i, x_j), then the flow leaving x_i at $\theta \in T$ will enter x_j at time period $\theta + \tau_{ij}$.

The task of the maximum dynamic flow finding has two main methods for its solving: with the time-expanded graph, proposed by L. Ford and D. Fulkerson and via the minimum cost flow algorithm applying, considering transit time along the arc as its cost and using the shortest path algorithm [3]. According to [2, 4] "time-expanded" fuzzy static graph $\widetilde{G}_p$ is the graph, constructed from the given fuzzy dynamic graph $\widetilde{G}$ by expanding the original dynamic graph in the time dimension by making a separate copy of every node $x_i \in X$ at every time $\theta \in T$. This copy is called "node-time pair" (x_i, θ).

Let $G_p = (X_p, A_p)$ represent fuzzy time-expanded static graph of the original dynamic fuzzy graph. The set of nodes X_p of the graph $\widetilde{G}_p$ is defined as $X_p = \{(x_i, \theta) : (x_i, \theta) \in X \times T\}$. The set of arcs A_p consists of arcs from each node-time pair $(x_i, \theta) \in X_p$ to every node-time pair $(x_j, \theta + \tau_{ij})$ where $x_j \in \Gamma(x_i)$ and $\theta + \tau_{ij}(\theta) \leq p$. Fuzzy arc capacities joining (x_i, θ) with $(x_j, \theta + \tau_{ij})$ are equal to u_{ij}.

The approach operating with the time-expanded network is more widespread despite the density of the second method. This is true due to the fact that algorithm based on flow decomposition doesn't consider dynamic structure of the transportation network (only flow transit times are taken into account), while the arc capacities and flow transit times are defined by constants. This fact allows to consider the model proposed by Ford-Fulkerson as "stationary-dynamic" one [5]. In the real life parameter of flow departure is crucial, because it influences the arc capacities and flow transit times of the network. For example, in the morning roads are not loaded, so they have high capacities, and therefore flow transit times are small. In the evening the roads are loaded, hence, their capacities are small and flow transit times become large. Consequently, we come to notion of "dynamic network", i.e. such a network, which parameters can vary over time. Consider fuzzy nature of network's arc capacities, mentioned in the first chapter. Therefore, it is necessary to propose methods for flows finding in fuzzy dynamic networks.

It is important to give definitions of the fuzzy "stationary-dynamic" (Definition 3.1) and dynamic network (Definition 3.2) [5].

Definition 3.1 *Fuzzy stationary-dynamic network is* [5] a network in the form of the fuzzy directed graph $\widetilde{G} = (X, \widetilde{A})$, where $X = \{x_1, x_2, \ldots, x_n\}$—the set of nodes, $\widetilde{A} = \{\langle \mu_{\widetilde{A}} \langle x_i, x_j \rangle / \langle x_i, x_j \rangle \rangle\}$, $\langle x_i, x_j \rangle \in X^2$, $\mu_{\widetilde{A}} \langle x_i, x_j \rangle$—fuzzy set of arcs, where $\mu_{\widetilde{A}}(x_i, x_j)$ is the membership degree of the directed arc $\langle x_i, x_j \rangle$ to the fuzzy set of directed arcs $\widetilde{A}$. Fuzzy arc capacity $\widetilde{u}_{ij}$ is used as $\mu_{\widetilde{A}}(x_i, x_j)$. Transit time τ_{ij} is set to each arc of the graph. There is time horizon $T = \{0, 1, \ldots, p\}$, which determines, that all flow units sent from the source must arrive at the sink not later, than in the time p. τ_{ij} is a positive number.

Definition 3.2 *Fuzzy dynamic network is* [5] a network $\widetilde{G} = (X, \widetilde{A})$, where $X = \{x_1, x_2, \ldots, x_n\}$—the set of nodes, $\widetilde{A} = \{\langle \mu_{\widetilde{A}} \langle x_i, x_j \rangle / \langle x_i, x_j \rangle \rangle\}$, $\langle x_i, x_j \rangle \in X^2$, $\mu_{\widetilde{A}} \langle x_i, x_j \rangle$—fuzzy set of arcs, where $\mu_{\widetilde{A}}(x_i, x_j)$ is the membership degree of the directed arc $\langle x_i, x_j \rangle$ to the fuzzy set of directed arcs $\widetilde{A}$. Transit arc capacity $\tilde{u}_{ij}(\theta)$ and transit time $\tau_{ij}(\theta)$, where $\theta \in T = \{0, 1, \ldots, p\}$—the flow departure time, are set to each arc of the graph. There is time horizon $T = \{0, 1, \ldots, p\}$, which determines, that all flow units sent from the source must arrive at the sink not later, than in the time p. τ_{ij} is a positive number.

Thus, tasks, considered in the present monograph, will rely on the notion of the dynamic network instead of "stationary-dynamic", observed in the conventional dynamic flow literature. As considered tasks will contain parameter of transit time along the arc, represented a fuzzy form, turn to the solution of these tasks in terms of partial uncertainty.

3.2 Maximum Flow Finding in Dynamic Network with Fuzzy Transit Arc Capacities

Let us consider the task of the maximum flow determining in the dynamic network, represented in the form of the fuzzy directed graph. The peculiarity of this task is that arc capacities are presented as fuzzy numbers. As arc capacities, parameters of transit times along the arcs depend on the flow departure time, therefore, we turn to the following problem statement [6]:

$$Maximize \; \tilde{v}(p), \tag{3.1}$$

$$\sum_{\theta=0}^{p} \left(\sum_{x_j \in \Gamma(x_i)} \tilde{\xi}_{ij}(\theta) - \sum_{x_j \in \Gamma^{-1}(x_i)} \tilde{\xi}_{ji}(\theta - \tau_{ji}(\theta)) \right) = \tilde{v}(p), \quad x_i = s, \tag{3.2}$$

$$\sum_{\theta=0}^{p} \left(\sum_{x_j \in \Gamma(x_i)} \tilde{\xi}_{ij}(\theta) - \sum_{x_j \in \Gamma^{-1}(x_i)} \tilde{\xi}_{ji}(\theta - \tau_{ji}(\theta)) \right) = \tilde{0}, \quad x_i \neq s, t; \; \theta \in T, \tag{3.3}$$

$$\sum_{\theta=0}^{p} \left(\sum_{x_j \in \Gamma(x_i)} \tilde{\xi}_{ij}(\theta) - \sum_{x_j \in \Gamma^{-1}(x_i)} \tilde{\xi}_{ji}(\theta - \tau_{ji}(\theta)) \right) = -\tilde{v}(p), \quad x_i = t, \tag{3.4}$$

$$\tilde{0} \leq \tilde{\xi}_{ij}(\theta) \leq \tilde{u}_{ij}(\theta), \quad \forall (x_i, x_j) \in \widetilde{A}, \; \theta \in T. \tag{3.5}$$

Equation (3.1) means that we maximize fuzzy amount of flow $\tilde{v}$ for p periods of time, i.e. $\tilde{v}(p)$. Equation (3.2) indicates that the maximum flow value $\tilde{v}$ for p time periods is equal to the flow, leaving the source for p time periods

$\sum_{\theta=0}^{p} \sum_{x_j \in \Gamma(x_i)} \tilde{\xi}_{ij}(\theta), x_i = s$. Equation (3.4) reflects that the maximum flow value $\tilde{v}$ for p time periods is equal to the total flow entering the sink for p time periods $\sum_{\theta=0}^{p} \sum_{x_j \in \Gamma^{-1}(x_i)} \tilde{\xi}_{ji}(\theta - \tau_{ji}(\theta)), x_i = t$. The total amount of flow entering the source $\sum_{\theta=0}^{p} \sum_{x_j \in \Gamma^{-1}(x_i)} \tilde{\xi}_{ji}(\theta - \tau_{ji}(\theta)), x_i = s$ for p time periods is equal to the total flow $\sum_{\theta=0}^{p} \sum_{x_j \in \Gamma(x_i)} \tilde{\xi}_{ij}(\theta), x_i = t$ leaving the sink for p time periods and is equal to $\tilde{0}$. For each node x_i except the source and the sink and for each time period θ the amount of flow $\sum_{\theta=0}^{p} \sum_{x_j \in \Gamma^{-1}(x_i)} \tilde{\xi}_{ji}(\theta - \tau_{ji}(\theta)), x_i \neq s, t$ entering x_i at each period of time $(\theta - \tau_{ji}(\theta))$ is equal to the total amount of flow $\sum_{\theta=0}^{p} \sum_{x_j \in \Gamma(x_i)} \tilde{\xi}_{ij}(\theta), x_i \neq s, t$ leaving x_i at time θ for p time periods as stated in (3.3). The inequality (3.5) indicates that the flows $\tilde{\xi}_{ij}(\theta)$ for all tome periods should be less than arc capacities of the corresponding arcs $\tilde{u}_{ij}(\theta)$ in the same time periods.

Time-expanding the initial fuzzy dynamic graph, introduce "time-expanding" static fuzzy graph, that is reflected in the *rule 3.1*, presented in [6]. Build fuzzy residual network according to the *rule 3.2*, presented in [6] for the maximum flow finding in the constructed graph. We will find the maximum flow in the fuzzy residual network corresponded to the static "time-expanded" graph via J. Edmonds and R. Karp, modified for using in fuzzy numbers.

Rule 3.1 of constructing the time-expanded fuzzy graph for the maximum flow finding in the dynamic network with fuzzy arc capacities [6].

Go to the time-expanded fuzzy static graph $\widetilde{G}_p$ from the given fuzzy dynamic graph $\widetilde{G}$ by expanding the original dynamic graph in the time dimension by making a separate copy of every node $x_i \in X$ at every time $\theta \in T$. Let $\widetilde{G}_p = (X_p, \widetilde{A}_p)$ represent fuzzy time-expanded static graph of the original dynamic fuzzy graph. The set of nodes X_p of the graph $\widetilde{G}_p$ is defined as $X_p = \{(x_i, \theta) : (x_i, \theta) \in X \times T\}$ The set of arcs $\widetilde{A}_p$ consists of arcs from each node-time pair $(x_i, \theta) \in X_p$ to every node-time pair $(x_j, \vartheta = \theta + \tau_{ij}(\theta))$, where $x_j \in \Gamma(x_i)$ and $\theta + \tau_{ij}(\theta) \leq p$. Fuzzy arc capacities $\tilde{u}(x_i, x_j, \theta, \vartheta)$ joining (x_i, θ) with (x_j, ϑ) are equal to $\tilde{u}_{ij}(\theta)$. Transit times $\tau(x_i, x_j, \theta, \vartheta)$ joining (x_i, θ) with (x_j, ϑ) are equal to $\tau_{ij}(\theta)$.

Rule 3.2 of constructing the fuzzy residual network of the "time-expanded" graph for the maximum flow finding in the dynamic network with fuzzy arc capacities [6].

Fuzzy residual network $\widetilde{G}_p^{\mu} = (X_p^{\mu}, \widetilde{A}_p^{\mu})$ constructed according to the "time-expanded" graph $\widetilde{G}_p = (X_p, \widetilde{A}_p)$ according to the flow values $\tilde{\xi}(x_i, x_j, \theta, \vartheta = \theta + \tau_{ij}(\theta))$ going from the nodes (x_i^{μ}, θ) to the nodes (x_j^{μ}, ϑ): each arc in the residual network $\widetilde{G}_p^{\mu}$, connected the "node-tome" pair (x_i^{μ}, θ) and (x_j^{μ}, ϑ) with the flow $\tilde{\xi}(x_i, x_j, \theta, \vartheta)$ sent at the time $\theta \in T$ has residual fuzzy arc capacity $\tilde{u}^{\mu}(x_i, x_j, \theta, \vartheta) = \tilde{u}(x_i, x_j, \theta, \vartheta) - \tilde{\xi}(x_i, x_j, \theta, \vartheta)$ with the transit time $\tau^{\mu}(x_i, x_j, \theta, \vartheta) = \tau(x_i, x_j, \theta, \vartheta)$ and reverse arc, connected (x_j^{μ}, ϑ) and (x_i^{μ}, θ) with fuzzy arc capacity $\tilde{u}^{\mu}(x_j, x_i, \vartheta, \theta) = \tilde{\xi}(x_i, x_j, \theta, \vartheta)$ and transit time $\tau^{\mu}(x_j, x_i, \vartheta), \theta = -\tau(x_i, x_j, \theta, \vartheta)$.

Represent algorithm according to the introduced rules.

Algorithm of the maximum flow finding in the dynamic network with fuzzy transit arc capacities.

Let us represent formal algorithm, describing the solution of the maximum flow finding in the dynamic network with fuzzy transit arc capacities and crisp transit times [6].

Step 1. Go to the time-expanded fuzzy static graph $\widetilde{G}_p$ from the given fuzzy dynamic graph $\widetilde{G}$ by expanding the original dynamic graph in the time dimension by making a separate copy of every node $x_i \in X$ at every time $\theta \in T$. According to the *rule 3.1* it is necessary to calculate the maximum dynamic flow leaving the group of sources, expanded for p periods and entering the group of sinks expanded for p periods no later than p. Let us introduce an artificial source s' and sink t' and join s' and t' by arcs with each true source and sink, respectively. Artificial arcs coming from and entering artificial nodes have infinite capacities.

Step 2. Build a fuzzy residual network $\widetilde{G}_p^\mu$ depending on the flow values going along the arcs of the graph $\widetilde{G}_p$. Fuzzy residual network $\widetilde{G}_p^\mu = (X_p^\mu, \widetilde{A}_p^\mu)$ constructed according to the "time-expanded" graph $\widetilde{G}_p = (X_p, \widetilde{A}_p)$ according to the flow values, starting with the zero flows according to the *rule 3.2*. Initially residual network coincides with the "time-expanded" static graph (as $\widetilde{\xi}(x_i, x_j, \theta, \vartheta) = \widetilde{0}$).

Step 3. Find the augmenting shortest path $\widetilde{P}_p^\mu$ in terms of the number of arcs, from the artificial source s' to the artificial sink t' in the constructed fuzzy residual network $\widetilde{G}_p^\mu$. The choice of the shortest path is according to breadth-first search.

3.1. Go to the **step 4** if the augmenting path $\widetilde{P}_p^\mu$ is found.

3.2. Fuzzy maximum flow value is $\widetilde{\xi}(x_i, x_j, \theta, \vartheta) + \widetilde{\delta}_p^\mu \times \widetilde{P}_p^\mu = \widetilde{v}(p)$ in the time-expanded fuzzy static graph if the path is failed to find from s' to t', then go to the **step 6**.

Step 4. Pass the maximum amount of flow units depending on the arc with minimal fuzzy residual capacity, $\widetilde{\delta}_p^\mu = \min[\widetilde{u}(\widetilde{P}_p^\mu)], \widetilde{u}(\widetilde{P}_p^\mu) = \min[\widetilde{u}^\mu(x_i, x_j, \theta, \vartheta)], (x_i, \theta), (x_j, \vartheta) \in \widetilde{P}_p^\mu$ along the found path.

Step 5. Update the fuzzy flow values in the graph $\widetilde{G}_p$: replace the fuzzy flow $\widetilde{\xi}(x_j, x_i, \theta, \vartheta)$ by $\widetilde{\xi}(x_j, x_i, \theta, \vartheta) - \widetilde{\delta}_p^\mu$ for arcs connecting node-time pair (x_i^μ, ϑ) with (x_j^μ, θ), such as, $((x_i^\mu, \vartheta), (x_j^\mu, \theta)) \notin \widetilde{A}_p, ((x_i^\mu, \vartheta), (x_j^\mu, 0)) \in \widetilde{A}_p^\mu$ in $\widetilde{G}_p^\mu$ along the corresponding arcs going from (x_j, θ) to (x_i, ϑ) from $\widetilde{G}_p$ and replace the fuzzy flow $\widetilde{\xi}(x_i, x_j, \theta, \vartheta)$ by $\widetilde{\xi}(x_i, x_j, \theta, \vartheta) + \widetilde{\delta}_p^\mu$ for arcs connecting node-time pair $(\widetilde{x}_i^\mu, \theta)$ with (x_j^μ, ϑ), such as $((x_i^\mu, \vartheta), (x_j^\mu, \theta)) \in \widetilde{A}_p^\mu$ in $\widetilde{G}_p^\mu$ along the arcs going from (x_i, θ) to (x_j, ϑ) from $\widetilde{G}_p$. Replace the fuzzy flow value in $\widetilde{G}_p$: $\widetilde{\xi}(x_i, x_j, \theta, \vartheta) \rightarrow \widetilde{\xi}(x_i, x_j, \theta, \vartheta) + \widetilde{\delta}_p^\mu \times \widetilde{P}_p^\mu$ and go to the **step 2**, starting with the updated flow values.

Step 6. Turn to the original dynamic fuzzy graph $\widetilde{G}$ if the maximum flow $\widetilde{\xi}(x_i, x_j, \theta, \vartheta) + \widetilde{\delta}_p^\mu \times \widetilde{P}_p^\mu = \widetilde{v}(p)$ is found in $\widetilde{G}_p$ from the artificial source s' to the artificial sink t' defined by the set of paths $\widetilde{P}_p^\mu$ such a way: cancel the artificial arcs and artificial nodes s' and t'. Therefore, the maximum flow of the value $\widetilde{v}(p)$ is obtained in the original graph, which is equivalent to the flow from the sources (initial node, expanded on the p periods) to the sinks (terminal node, expanded on the p periods) in $\widetilde{G}_p$ after cancelling artificial nodes. Each path connecting the node-time pairs (s, ς) and $(t, \psi = \varsigma + \tau_{st}(\varsigma)), \psi \in T$ which the flow $\widetilde{\xi}(s, t, \theta, \varsigma)$ goes along is equal to the flow $\widetilde{\xi}_{st}(\vartheta)$

Consider example, illustrating the implementation of the maximum flow finding algorithm in the dynamic network with fuzzy transit arc capacities.

Numerical example 5

Let the network, which is the part of the railway network, is represented in the form of the fuzzy directed graph, obtained from «Object Land», as shown in Fig. 3.1. Fuzzy arc capacities and transit times along the arcs which depend on the flow departure time are presented in Tables 3.1 и 3.2. The node x_1 is the source, the node x_7 is the sink. It is necessary to find the maximum flow in the fuzzy dynamic graph for 3 periods of time and represent it in the form of the fuzzy trapezoidal number, if there are basic values of arc capacities, represented in the fuzzy trapezoidal number, as shown in Fig. 2.4.

Fig. 3.1 Initial dynamic network $\widetilde{G}$

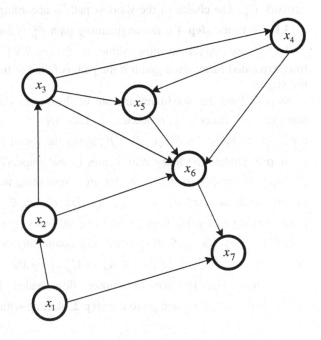

Table 3.1 Fuzzy arc capacities $\tilde{u}_{ij}$, which depend on the flow departure time θ

Arc capacities	Fuzzy arc capacities $\tilde{u}_{ij}$ at the time θ, units			
	0	1	2	3
(x_1, x_2)	$[\widetilde{10,14}]$	$[\widetilde{8,11}]$	$[\widetilde{8,11}]$	$[\widetilde{10,14}]$
(x_1, x_7)	$[\widetilde{15,21}]$	$[\widetilde{10,14}]$	$[\widetilde{15,21}]$	$[\widetilde{12,18}]$
(x_2, x_3)	$[\widetilde{18,24}]$	$[\widetilde{15,21}]$	$[\widetilde{18,24}]$	$[\widetilde{18,24}]$
(x_2, x_6)	$[\widetilde{25,34}]$	$[\widetilde{20,27}]$	$[\widetilde{25,34}]$	$[\widetilde{18,24}]$
(x_3, x_4)	$[\widetilde{30,40}]$	$[\widetilde{25,34}]$	$[\widetilde{25,34}]$	$[\widetilde{30,40}]$
(x_3, x_5)	$[\widetilde{25,34}]$	$[\widetilde{25,34}]$	$[\widetilde{20,27}]$	$[\widetilde{20,27}]$
(x_3, x_6)	$[\widetilde{40,53}]$	$[\widetilde{30,40}]$	$[\widetilde{30,40}]$	$[\widetilde{25,34}]$
(x_4, x_5)	$[\widetilde{30,40}]$	$[\widetilde{25,34}]$	$[\widetilde{18,24}]$	$[\widetilde{25,34}]$
(x_4, x_6)	$[\widetilde{18,24}]$	$[\widetilde{18,24}]$	$[\widetilde{15,21}]$	$[\widetilde{30,40}]$
(x_5, x_6)	$[\widetilde{20,27}]$	$[\widetilde{18,24}]$	$[\widetilde{15,21}]$	$[\widetilde{18,24}]$
(x_6, x_7)	$[\widetilde{30,40}]$	$[\widetilde{18,24}]$	$[\widetilde{12,18}]$	$[\widetilde{25,34}]$

Table 3.2 Parameters of transit time τ_{ij} along the arcs, which depend on the departure time θ

Arcs of the graph	Transit time along the arc τ_{ij} at time periods θ, units of time.			
	0	1	2	3
(x_1, x_2)	1	2	3	2
(x_1, x_7)	1	2	2	1
(x_2, x_3)	3	3	3	1
(x_2, x_6)	4	1	3	2
(x_3, x_4)	2	3	3	2
(x_3, x_5)	5	3	2	3
(x_3, x_6)	4	3	2	2
(x_4, x_5)	2	2	3	1
(x_4, x_6)	1	3	3	2
(x_5, x_6)	2	3	3	2
(x_6, x_7)	4	2	1	1

Step 1. Turn to the "time-expanded" static graph from the initial dynamic graph according to the *rule 3.1*, that represented in Fig. 3.2.

Step 2. Fuzzy residual network $\widetilde{G}_p^\mu$ coincides with the initial network $\widetilde{G}_p$, represented in Fig. 3.2.

Step 3. Find the first augmenting path $\widetilde{P}_1^\mu$ according to the breadth-first-search in $\widetilde{G}_p^\mu$: $\widetilde{P}_1^\mu = s', (x_1, 0), (x_7, 1), t'$.

Step 4. Pass min $[\infty, [\widetilde{15,21}], \infty]$, i.e. $[\widetilde{15,21}]$ flow units along the path $\widetilde{P}_1^\mu = s', (x_1, 0), (x_7, 1), t'$.

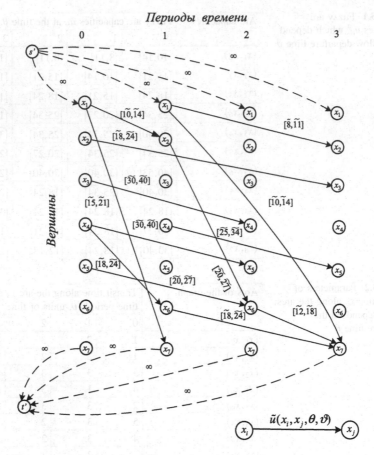

Fig. 3.2 $\widetilde{G}_p$—«time-expanded» graph $\widetilde{G}$

Step 5. Update the flow values in the graph $\widetilde{G}_p$. The flow $\tilde{\xi}(x_i, x_j, \theta, \vartheta) = [\tilde{0}, \tilde{0}]$ turns to $\tilde{\xi}(x_i, x_j, \theta, \vartheta) = [\tilde{0}, \tilde{0}] + [1\tilde{5}, 2\tilde{1}] = [1\tilde{5}, 2\tilde{1}]$. Build graph with the new flow value, as shown in Fig. 3.3 and turn to the **step 2**.

Step 2. Determine arc capacities of fuzzy residual network $\widetilde{G}_p^\mu$ according to the *rule 3.2 for* the graph in Fig. 3.3 according to the arc flow values.

Build fuzzy residual network, as shown in Fig. 3.4.

Step 3. Find the second shortest path $\widetilde{P}_2^\mu$ according to the breadth-first-search in $\widetilde{G}_p^\mu$: $\widetilde{P}_2^\mu = s', (x_1, 1), (x_7, 3), t'$.

Step 4. Pass min $[\infty, [1\tilde{0}, 1\tilde{4}], \infty]$, i.e. $[1\tilde{0}, 1\tilde{4}]$ flow units along the path $\widetilde{P}_2^\mu = s', (x_1, 1), (x_7, 3), t'$.

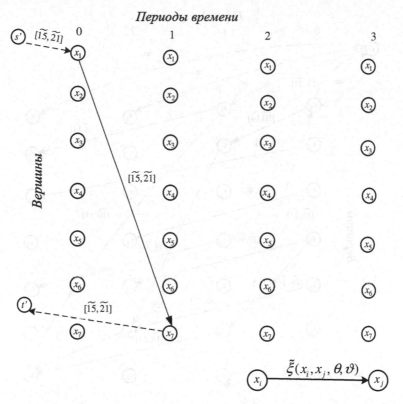

Fig. 3.3 Graph $\widetilde{G}_p$ with the flow $[\widetilde{15}, \widetilde{21}]$ units

Step 5. Update the flow value in $\widetilde{G}_p$:

The flow $\widetilde{\xi}(x_i, x_j, \theta, \vartheta) = [\widetilde{15}, \widetilde{21}]$ turns to $\widetilde{\xi}(x_i, x_j, \theta, \vartheta) = [\widetilde{15}, \widetilde{21}] + [\widetilde{10}, \widetilde{14}] = [\widetilde{25}, \widetilde{35}]$. Build a graph with the new flow value, as shown in Fig. 3.5 and turn to the step 2.

Step 2. Determine arc capacities of the fuzzy residual network $\widetilde{G}_p^\mu$ according to the *rule 3.2* for the graph in Fig. 3.5 according to the flow values, passing along the arcs of the graph.

Build fuzzy residual network, as shown in Fig. 3.6.

Step 3. Find the third shortest path $\widetilde{P}_3^\mu$ according to the breadth-first-search in the residual network $\widetilde{G}_p^\mu$: $\widetilde{P}_3^\mu = s', (x_1, 0), (x_2, 1), (x_6, 2), (x_7, 3), t'$.

Step 4. Pass min $[\infty, [\widetilde{10}, \widetilde{14}], [\widetilde{20}, \widetilde{27}], [\widetilde{12}, \widetilde{18}], \infty]$, i.e. $[\widetilde{10}, \widetilde{14}]$ flow units along the path $\widetilde{P}_3^\mu = s', (x_1, 0), (x_2, 1), (x_6, 2), (x_7, 3), t'$.

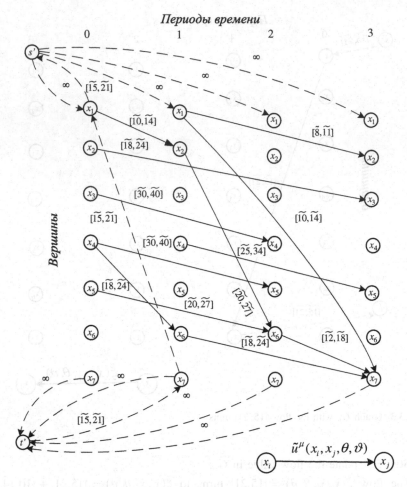

Fig. 3.4 Fuzzy residual network $\widetilde{G}_p^\mu$ for the graph in Fig. 3.3

Step 5. Update flow values in $\widetilde{G}_p$:

Flow $\widetilde{\xi}(x_i, x_j, \theta, \vartheta) = [2\widetilde{5}, 3\widetilde{5}]$ turns to $\widetilde{\xi}(x_i, x_j, \theta, \vartheta) = [2\widetilde{5}, 3\widetilde{5}] + [1\widetilde{0}, 1\widetilde{4}] = [3\widetilde{5}, 4\widetilde{9}]$. Build a graph with the new flow value, as shown in Fig. 3.7 and turn to the step 2.

Step 2. Determine arc capacities of the fuzzy residual network $\widetilde{G}_p^\mu$ according to the *rule 3.2 f*or the graph in Fig. 3.7 according to the flow values in the graph.

Build fuzzy residual network, as shown in Fig. 3.8.

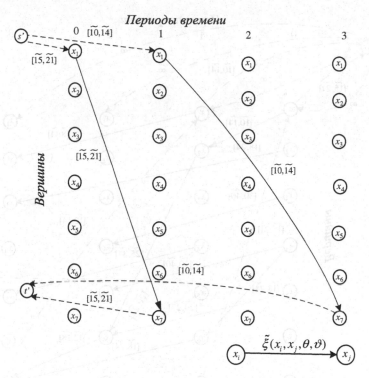

Fig. 3.5 Graph $\widetilde{G}_p$ with the flow of $[2\widetilde{5}, 3\widetilde{5}]$ units

Step 3. There is no augmenting path from s' to t', therefore, the maximum flow in $\widetilde{G}_p$, presented in Fig. 3.9 equals to $[3\widetilde{5}, 4\widetilde{9}]$ flow units.

Step 6. As the maximum flow in the "time-expanded" graph $\widetilde{G}_p$ is found, turn to the initial dynamic graph $\widetilde{G}$: the maximum flow of $[3\widetilde{5}, 4\widetilde{9}]$ units in $\widetilde{G}$ is defined by the flows: $x_1 \rightarrow x_7$, the flow $[1\widetilde{5}, 2\widetilde{1}]$ is passed along at the zero time period, $x_1 \rightarrow x_7$, the flow $[1\widetilde{0}, 1\widetilde{4}]$ is passed along at the first time period and $x_1 \rightarrow x_2 \rightarrow x_6 \rightarrow x_7$, the flow $[1\widetilde{0}, 1\widetilde{4}]$ is passed along at the zero time period.

Define borders of uncertainty of the obtained fuzzy interval $[3\widetilde{5}, 4\widetilde{9}]$, corresponded to the maximum flow in $\widetilde{G}$. The found result is between two basic values of arc capacities: $[2\widetilde{3}, 3\widetilde{1}]$ with the left deviation $l_1^L = 6$, right deviation—$l_1^R = 6$ and $[3\widetilde{9}, 5\widetilde{1}]$ with the left deviation $l_2^L = 8$, right deviation—$l_2^R = 10$. According to (2.5) we obtain:

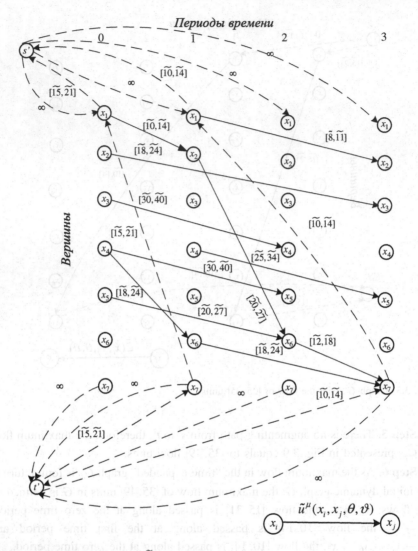

Fig. 3.6 Fuzzy residual network $\widetilde{G}_p^{\mu}$ for the graph in Fig. 3.5

$$l^L = \frac{(a_2 - a')}{(a_2 - a_1)} \times l_1^L + \left(1 - \frac{(a_2 - a')}{(a_2 - a_1)}\right) \times l_2^L$$

$$= \frac{(39 - 35)}{(39 - 23)} \times 6 + \left(1 - \frac{(39 - 35)}{(39 - 23)}\right) \times 8$$

$$= 7.5 \approx 8$$

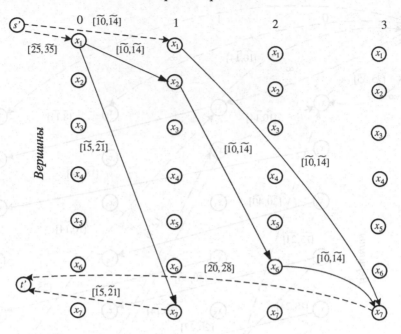

Периоды времени

Вершины

Fig. 3.7 Graph $\widetilde{G}_p$ with the flow of $[3\widetilde{5}, 4\widetilde{9}]$ units

$$
\begin{aligned}
l^R &= \frac{(a_2 - a')}{(a_2 - a_1)} \times l_1^R + \left(1 - \frac{(a_2 - a')}{(a_2 - a_1)}\right) \times l_2^R \\
&= \frac{(51 - 49)}{(51 - 31)} \times 6 + \left(1 - \frac{(51 - 49)}{(51 - 31)}\right) \times 10 \\
&= 9.6 \approx 10.
\end{aligned}
$$

Therefore, the maximum flow in the fuzzy network is obtained, which can be represented in the form of the fuzzy trapezoidal number of (35,49,8,10) units, as shown in Fig. 3.10.

According to Fig. 3.10 the maximum flow with the degree of confidence 0.8 will be within the interval $[3\widetilde{3}, 5\widetilde{1}]$, but anyway the maximum flow will be not less than 27 units and no more than 59 units.

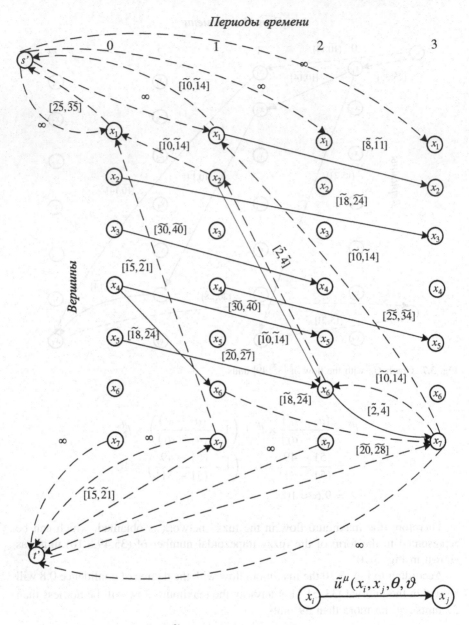

Fig. 3.8 Fuzzy residual network $\widetilde{G}_p^{\mu}$ for the graph in Fig. 3.7

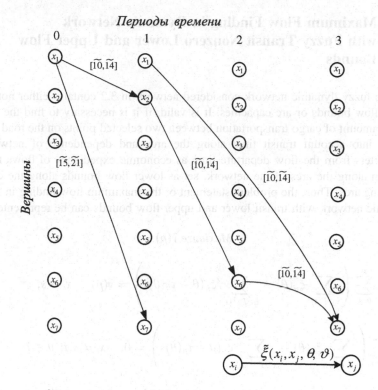

Fig. 3.9 Graph $\widetilde{G}_p$ with the flow of $[3\widetilde{5}, 4\widetilde{9}]$ units

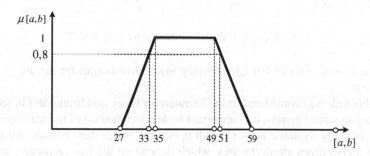

Fig. 3.10 Maximum flow in the form of the fuzzy residual number of (35,49,8,10) units

3.3 Maximum Flow Finding in Dynamic Network with Fuzzy Transit Nonzero Lower and Upper Flow Bounds

Let the fuzzy dynamic network considered network in 3.2 contains either nonzero lower flow bounds or arc capacities. It is valid, if it is necessary to find the maximum amount of cargo transportation between two selected points on the road map, taking into account transit time along the arcs and dependence of network's parameters from the flow departure time as economic expediency of flows transmission along the arcs of the network, set as lower flow bounds along the corresponding arcs. Thus, the problem statement of the maximum flow finding in fuzzy dynamic network with transit lower and upper flow bounds can be represented as:

$$Maximize\ \tilde{v}(p), \tag{3.6}$$

$$\sum_{\theta=0}^{p} \left(\sum_{x_j \in \Gamma(x_i)} \tilde{\xi}_{ij}(\theta) - \sum_{x_j \in \Gamma^{-1}(x_i)} \tilde{\xi}_{ji}(\theta - \tau_{ji}(\theta)) \right) = \tilde{v}(p), \quad x_i = s, \tag{3.7}$$

$$\sum_{\theta=0}^{p} \left(\sum_{x_j \in \Gamma(x_i)} \tilde{\xi}_{ij}(\theta) - \sum_{x_j \in \Gamma^{-1}(x_i)} \tilde{\xi}_{ji}(\theta - \tau_{ji}(\theta)) \right) = \tilde{0}, \quad x_i \neq s, t;\ \theta \in T, \tag{3.8}$$

$$\sum_{\theta=0}^{p} \left(\sum_{x_j \in \Gamma(x_i)} \tilde{\xi}_{ij}(\theta) - \sum_{x_j \in \Gamma^{-1}(x_i)} \tilde{\xi}_{ij}(\theta - \tau_{ij}(\theta)) \right) = -\tilde{v}(p), \quad x_i = t, \tag{3.9}$$

$$\tilde{l}_{ij}(\theta) \leq \tilde{\xi}_{ij}(\theta) \leq \tilde{u}_{ij}(\theta), \quad \theta + \tau_{ij}(\theta) \leq p,\ \theta \in T. \tag{3.10}$$

In the model (3.6)–(3.10) $\tilde{l}_{ij}(\theta)$—fuzzy lower flow bound for the arc (x_i, x_j) at the time θ.

As this task was considered in the literature in fuzzy conditions and in terms of stationary-dynamic graphs, it is important to develop method of the maximum flow finding in fuzzy dynamic network with upper and lower flow bounds and parameters of transit times along the arcs, which depend on the flow departure time.

To realize this algorithm it is necessary to introduce the rule of transition to the "time-expanded" static graph, corresponded to the initial dynamic graph, that reflected in the *rule 3.3*.

Rule 3.3 of constructing the "time-expanded" fuzzy graph for the maximum flow finding in dynamic network with fuzzy nonzero transit lower and upper flow.

Turn to the "time-expanded" for p periods fuzzy static graph $\tilde{G}_p$ from the initial fuzzy dynamic graph $\tilde{G} = (X, \tilde{A})$ expanding the original dynamic graph in the time dimension by making a separate copy of every node $x_i \in X$ at every time $\theta \in T$. Let

$\widetilde{G}_p = (X_p, \widetilde{A}_p)$ represent fuzzy time-expanded static graph of the original dynamic fuzzy graph. The set of nodes X_p of the graph $\widetilde{G}_p$ is defined as $X_p = \{(x_i, \theta) : (x_i, \theta) \in X \times T\}$. The set of arcs $\widetilde{A}_p$ consists of arcs from each node-time pair $(x_i, \theta) \in X_p$ to every node-time pair $(x_j, \vartheta = \theta + \tau_{ij}(\theta))$, where $x_j \in \Gamma(x_i)$ and $\theta + \tau_{ij}(\theta) \leq p$. Lower flow bounds $\widetilde{l}(x_i, x_j, \theta, \vartheta)$ joining (x_i, θ) with (x_j, ϑ) in $\widetilde{G}_p$ are equal to $\widetilde{l}_{ij}(\theta)$. Fuzzy upper flow bounds $\widetilde{u}(x_i, x_j, \theta, \vartheta)$ joining (x_i, θ) with (x_j, ϑ) are equal to $\widetilde{u}_{ij}(\theta)$. Transit times $\tau(x_i, x_j, \theta, \vartheta)$ joining (x_i, θ) with (x_j, ϑ) are equal to $\tau_{ij}(\theta)$.

Turn to the fuzzy "time-expanded" graph without lower flow bounds according to the *rule 3.4*.

Rule 3.4 of turning to the corresponding fuzzy graph without lower flow bounds from the "time-expanded" fuzzy graph for the maximum flow finding in dynamic network with fuzzy nonzero transit lower and upper flow bounds.

Turn to the fuzzy graph $\widetilde{G}_p^* = (X_p^*, \widetilde{A}_p^*)$ without lower flow bounds from the "time-expanded" fuzzy graph $\widetilde{G}_p = (X_p, \widetilde{A}_p)$. Introduce the artificial source s^* and sink t^* and arcs connecting the node-time pair $(t, \forall \theta \in T)$ and $(s, \forall \theta \in T)$ with upper fuzzy flow bound $\widetilde{u}^*(t, s, \forall \theta \in T, \forall \theta \in T) = \infty$, lower fuzzy flow bound $\widetilde{l}^*(t, s, \forall \theta \in T, \forall \theta \in T) = \widetilde{0}$ in the graph $\widetilde{G}_p$. It means that every node t at each time period from p is connected with every node s at all time periods in the graph $\widetilde{G}_p^*$ Introduce the following modification for each arc connecting the node-time pair (x_i, θ) with the node-time pair $(x_j, \vartheta = \theta + \tau_{ij}(\theta))$ with nonzero lower fuzzy flow bound $\widetilde{l}(x_i, x_j, \theta, \vartheta \neq \widetilde{0})$: (1) reduce $\widetilde{u}(x_i, x_j, \theta, \vartheta)$ to $\widetilde{u}^*(x_i, x_j, \theta, \vartheta) = \widetilde{u}(x_i, x_j, \theta, \vartheta) - \widetilde{l}(x_i, x_j, \theta, \vartheta), \widetilde{l}(x_i, x_j, \theta, \vartheta)$ to $\widetilde{0}$. (2) Introduce the arcs connecting s^* with (x_j, ϑ), and the arcs connecting t^* with (x_i, ϑ) with upper fuzzy flow bounds equal to lower fuzzy flow bounds $\widetilde{u}^*(s^*, x_j, \forall, \vartheta) = \widetilde{u}^*(x_i, t, \theta, \forall) = \widetilde{l}(x_i, x_j, \theta, \forall)$ zero lower fuzzy flow bounds $\widetilde{l}^*(s^*, x_j, \forall, \vartheta) = \widetilde{l}^*(x_i, t, \theta, \forall) = \widetilde{0}$.

Search the maximum flow in the fuzzy residual network, constructed according to the "time-expanded" graph without lower flow bounds. Introduce synthesized Definition 3.3 of the fuzzy residual network of the "time-expanded" graph, modified considering the structure of the "time-expanded" graph and fuzzy character of the upper and lower flow bounds.

Definition 3.3 Fuzzy residual network $\widetilde{G}_p^{*\mu} = (X_p^{*\mu}, \widetilde{A}_p^{*\mu})$ of the time-expanded graph for the maximum flow finding in the dynamic network with nonzero transit upper and lower flow bounds is the network without lower fuzzy flow bounds $\widetilde{G}_p^* = (X_p^*, \widetilde{A}_p^*)$, which is constructed due to the following rules: if the flow connecting the node-time pair (x_i^*, θ) with the node-time pair (x_j^*, ϑ) is less than arc capacity in $\widetilde{G}_p^*$, i.e. $\widetilde{\xi}^*(x_i, x_j, \theta, \vartheta) < \widetilde{u}^*(x_i, x_j, \theta, \vartheta)$, then include the corresponding arc, going from the node-time pair $(x_i^{*\mu}, \theta)$ to the node-time pair $(x_j^{*\mu}, \vartheta)$ in $\widetilde{G}_p^{*\mu}$ with arc capacity $\widetilde{u}^{*\mu}(x_i, x_j, \theta, \vartheta) = \widetilde{u}^*(x_i, x_j, \theta, \vartheta) - \widetilde{\xi}^*(x_i, x_j, \theta, \vartheta)$ and transit time $\tau^{*\mu}(x_i, x_j, \theta, \vartheta) = \tau^*(x_i, x_j, \theta, \vartheta)$. If the flow, going from the node-time pair (x_i^*, θ) to the node-time pair (x_j^*, ϑ) is more than $\widetilde{0}$ in $\widetilde{G}_p^*$, i.e. $\widetilde{\xi}^*(x_i, x_j, \theta, \vartheta) > \widetilde{0}$, then include the corresponding arc, going from the node-time pair $(x_j^{*\mu}, \vartheta)$ to the node-time pair $(x_i^{*\mu}, \theta)$ in $\widetilde{G}_p^{*\mu}$ with arc capacity $\tilde{u}^{*\mu}(x_j, x_i, \vartheta, \theta) = \tilde{\xi}^*(x_i, x_j, \theta, \vartheta)$ and transit time $\tau^{*\mu}(x_j, x_i, \vartheta, \theta) = -\tau^*(x_i, x_j, \theta, \vartheta)$.

The criteria of the presence of the feasible flow in the time-expanded fuzzy graph, corresponded to the initial dynamic graph, is the maximum flow obtaining, that is equal to the sum of the lower flow bounds in the modified graph without lower flow bounds. The reverse transition from the time-expanded fuzzy graph without lower flow bounds with the maximum flow to the time-expanded graph with the feasible flow, corresponded to the initial dynamic is made by the *rule 3.5*.

Rule 3.5 of transition from the time-expanded fuzzy graph without lower flow bounds with the found maximum flow to the graph with the feasible flow for the for the maximum flow finding in dynamic network with fuzzy nonzero transit lower and upper flow bounds.

Turn to the graph $\widetilde{G}_p$ from the graph $\widetilde{G}_p^*$ as following: reject artificial nodes and arcs, connecting them with other nodes. The feasible flow vector $\tilde{\xi} = (\tilde{\xi}(x_i, x_j, \theta, \vartheta))$ of the value $\tilde{\sigma}$ is defined as: $\tilde{\xi}(x_i, x_j, \theta, \vartheta) = \tilde{\xi}^*(x_i, x_j, \theta, \vartheta) + \tilde{l}(x_i, x_j, \theta, \vartheta)$, where $\tilde{\xi}^*(x_i, x_j, \theta, \vartheta)$—the flows, going along the arcs of the graph $\widetilde{G}_p^*$ after deleting all artificial nodes and connecting arcs.

If the feasible flow in the "time-expanded" fuzzy graph is found, modify it, finding the maximum flow, according to the *rule 3.6*. If the maximum flow equals the sum of the lower flow bounds wasn't found in the "time-expanded" graph without lower flow bounds, therefore, there is no feasible flow in this graph and, therefore, the task has no solution.

Rule 3.6 of the fuzzy residual network constructing with the feasible flow vector for the maximum flow finding in dynamic network with fuzzy nonzero transit lower and upper flow bounds for all arcs, if $\tilde{\xi}(x_i, x_j, \vartheta, \theta) < \tilde{u}(x_i, x_j, \vartheta, \theta)$, then include the corresponding arc in $G(\tilde{\xi}^\mu(x_i, x_j, \vartheta, \theta))$ with the arc capacity $\tilde{u}^\mu(x_i, x_j, \vartheta, \theta) = \tilde{u}(x_i, x_j, \vartheta, \theta) - \tilde{\xi}(x_i, x_j, \theta, \vartheta)$. For all arcs, if $\tilde{\xi}(x_i, x_j, \vartheta, \theta) > \tilde{l}(x_i, x_j, \vartheta, \theta)$, then include the corresponding arc in $G(\tilde{\xi}^\mu(x_i, x_j, \vartheta, \theta))$ with the arc capacity $\tilde{u}^\mu(x_j, x_i, \theta, \vartheta) = \tilde{\xi}(x_i, x_j, \vartheta, \theta) - \tilde{l}(x_i, x_j, \vartheta, \theta)$.

If the flow, going from the node-time pair (x_i, θ) to the node-time pair (x_j, ϑ) is less, than arc capacity in $\widetilde{G}_p$, i.e. $\tilde{\xi}(x_i, x_j, \theta, \vartheta) < \tilde{u}(x_i, x_j, \theta, \vartheta)$, then include the corresponding arc (x_i^μ, θ) from the node-time pair to the node-time pair (x_j^μ, ϑ) in $\widetilde{G}_p^\mu(\tilde{\xi})$ with arc capacity $\tilde{u}^\mu(x_i, x_j, \theta, \vartheta) = \tilde{u}(x_i, x_j, \theta, \vartheta) - \tilde{\xi}(x_i, x_j, \theta, \vartheta)$ and transit time $\tau^\mu(x_i, x_j, \theta, \vartheta) = \tau(x_i, x_j, \theta, \vartheta)$. If the flow, going from the node-time pair (x_i, θ) to the node-time pair (x_j, ϑ) is more than the lower flow bound in $\widetilde{G}_p$, i.e. $\tilde{\xi}(x_i, x_j, \theta, \vartheta) = \tilde{l}(x_i, x_j, \theta, \vartheta)$, then include the corresponding arc, going from the node-time pair (x_j^μ, ϑ) to the node-time pair (x_i^μ, θ) in $\widetilde{G}_p^\mu(\tilde{\xi})$ with arc capacity $\tilde{u}^\mu(x_j, x_i, \vartheta, \theta) = \tilde{\xi}(x_i, x_j, \theta, \vartheta) - \tilde{l}(x_i, x_j, \vartheta, \theta)$ and transit time $\tau^\mu(x_j, x_i, \vartheta, \theta) = \tau(x_i, x_j, \theta, \vartheta)$.

Based on the formulated rules and definitions turn to the solving of the maximum flow finding with nonzero lower flow bounds in dynamic network in terms of partial uncertainty.

Algorithm of the maximum flow finding with transit nonzero lower and upper flow bounds in fuzzy dynamic network.

Let us represent algorithm of the maximum flow finding in fuzzy dynamic network with transit nonzero lower and upper flow bounds.

Step 1. Go to the time-expanded fuzzy static graph $\widetilde{G}_p$ from the given fuzzy dynamic graph $\widetilde{G}$ by expanding the original dynamic graph in the time dimension by making a separate copy of every node $x_i \in X$ at every time $\theta \in T$ according to the *rule 3.3*.

Step 2. Determine, if the time-expanded fuzzy graph $\widetilde{G}_p$, corresponding to the initial dynamic graph $\widetilde{G}$, has a feasible flow. Turn to the graph $\widetilde{G}_p^* = (X_p^*, \widetilde{A}_p^*)$ according to the *rule 3.4*.

Step 3. Build a fuzzy residual network $\widetilde{G}_p^{*\mu}$ depending on the flow values going along the arcs of the graph $\widetilde{G}_p^*$ due to the Definition 3.3.

Step 4. Search the augmenting shortest path (in terms of the number of arcs) $\widetilde{P}_p^{*\mu}$ from the artificial source s^* to the artificial sink t^* in the constructed fuzzy residual network according to the breadth-first-search.

(I) Go to the step 5 if the augmenting path $\widetilde{P}_p^{*\mu}$ is found.

(II) The flow value $\widetilde{\phi}^* < \sum_{\widetilde{l}(x_j, x_i, \vartheta, \theta) \neq \widetilde{0}} \widetilde{l}(x_i, x_j, \theta, \vartheta)$ is obtained, which is the maximum flow in $\widetilde{G}_p^*$, if the path is failed to find. It means that it is impossible to pass any unit of flow, but not all the artificial arcs are saturated. Therefore, the time-expanded graph $\widetilde{G}_p$ has no feasible flow as the initial dynamic fuzzy graph $\widetilde{G}$ and the task has no solution. Exit.

Step 5. Pass the minimum from the arc capacities $\widetilde{\delta}_p^{*\mu} = \min[\widetilde{u}(\widetilde{P}_p^{*\mu})], \widetilde{u}(\widetilde{P}_p^{*\mu}) = \min[\widetilde{u}^{*\mu}(x_i, x_j, \theta, \vartheta), (x_i, \theta), (x_j, \vartheta) \in \widetilde{P}_p^{*\mu}$ along this path $\widetilde{P}_p^{*\mu}$.

Step 6. Update the fuzzy flow values in the graph $\widetilde{G}_p^*$: replace the fuzzy flow $\widetilde{\xi}^*(x_j, x_i, \theta, \vartheta)$ along the corresponding arcs going from (x_j^*, θ) to (x_i^*, ϑ) from $\widetilde{G}_p^*$ by $\widetilde{\xi}^*(x_j, x_i, \theta, \vartheta) - \widetilde{\delta}_p^{*\mu}$ for arcs connecting node-time pair $(x_i^{*\mu}, \vartheta)$ with $(x_j^{*\mu}, \theta)$ in $\widetilde{G}_p^{*\mu}$, such as $((x_i^{*\mu}, \vartheta), (x_j^{*\mu}, \theta)) \notin \widetilde{A}_p^*$, $((x_i^{*\mu}, \vartheta), (x_j^{*\mu}, \theta)) \in \widetilde{A}_p^{*\mu}$ and replace the fuzzy flow $\widetilde{\xi}^*(x_i, x_j, \theta, \vartheta)$ along the arcs going from (x_i^*, θ) to (x_j^*, ϑ) from $\widetilde{G}_p^*$ by $\widetilde{\xi}^*(x_i, x_j, \theta, \vartheta) + \widetilde{\delta}_p^{*\mu}$ for arcs connecting node-time pair $(x_i^{*\mu}, \theta)$ with $(x_j^{*\mu}, \vartheta)$ in $\widetilde{G}_p^{*\mu}$, such as $((x_i^{*\mu}, \theta), (x_j^{*\mu}, \vartheta)) \in \widetilde{A}_p^*$, $((x_i^{*\mu}, \theta), (x_j^{*\mu}, \vartheta)) \in \widetilde{A}_p^{*\mu}$. Replace $\widetilde{\xi}^*(x_i, x_j, \theta, \vartheta)$ by $\widetilde{\xi}^*(x_i, x_j, \theta, \vartheta) + \widetilde{\delta}_p^{*\mu} \widetilde{P}_p^{*\mu}$.

Step 7. Compare flow value $\widetilde{\xi}^*(x_i, x_j, \theta, \vartheta) + \widetilde{\delta}_p^{*\mu} \widetilde{P}_p^{*\mu}$ and $\sum_{\widetilde{l}(x_j, x_i, \vartheta, \theta) \neq \widetilde{0}}$ $\widetilde{l}(x_i, x_j, \theta, \vartheta)$:

7.1. If the flow value $\tilde{\xi}^*(x_i, x_j, \theta, \vartheta) + \tilde{\delta}_p^{*\mu} \tilde{P}_p^{*\mu}$ is less than $\sum_{\tilde{l}(x_j, x_i, \vartheta, \theta) \neq \tilde{0}}$ $\tilde{l}(x_i, x_j, \theta, \vartheta)$, i.e. not all artificial arcs become saturated, go to the **step 3**.

7.2. If the flow value $\tilde{\xi}^*(x_i, x_j, \theta, \vartheta) + \tilde{\delta}_p^{*\mu} \tilde{P}_p^{*\mu}$ is equal to $\sum_{\tilde{l}(x_j, x_i, \vartheta, \theta) \neq \tilde{0}}$ $\tilde{l}(x_i, x_j, \theta, \vartheta)$, i.e. all arcs from the artificial source to the artificial sink become saturated, then the value $\tilde{\xi}^*(x_i, x_j, \theta, \vartheta) + \tilde{\delta}_p^{*\mu} \tilde{P}_p^{*\mu}$ is required value of maximum flow $\tilde{\sigma}^*$ In this case the total flow along the artificial arcs connecting the node-time pairs $(t, \forall \theta \in T)$ with $(s, \forall \theta \in T)$, which is equal to $\sum_{\theta=0}^{p} \tilde{\xi}^*(t, s, \forall \theta \in T, \forall \theta \in T)$ in $\tilde{G}_p^*$ determines the feasible flow in time-expanded graph $\tilde{G}_p$ with the flow value $\sum_{\theta=0}^{p} \tilde{\xi}^*(t, s, \forall \theta \in T, \forall \theta \in T) = \tilde{\sigma}$. Turn to the graph $\tilde{G}_p$ from the graph $\tilde{G}_p^*$ according to the *rule 3.5*. The network $\tilde{G}_p(\tilde{\xi})$ is obtained. Go to the **step 8**.

Step 8. Construct the residual network $\tilde{G}_p^{\mu}(\tilde{\xi})$ taking into account the feasible flow vector $\tilde{\xi} = (\tilde{\xi}(x_i, x_j, \theta, \vartheta))$ in $\tilde{G}_p(\tilde{\xi})$ adding the artificial source and sink and the arcs with infinite arc capacity, connecting s' with true sources and t' with true sinks according to the *rule 3.6*.

Step 9. Define the shortest path $\tilde{P}_p^{\mu}$ according to the breadth-first-search from s' to t' in the constructed residual network $\tilde{G}_p^{\mu}(\tilde{\xi})$

(I)   Go to the **step 10** if the augmenting path $\tilde{P}_p^{\mu}$ is found.

(II) The maximum flow $\tilde{\xi}(x_i, x_j, \theta, \vartheta) + \tilde{\delta}_p^{\mu} \tilde{P}_p^{\mu} = \tilde{v}(p)$ in $\tilde{G}_p(\tilde{\xi})$ is found if the path is failed to find, then the maximum flow in "time-expanded" static fuzzy graph can be found at the **step 12**.

Step 10. Pass the flow value $\tilde{\delta}_p^{\mu} = \min[\tilde{u}(\tilde{P}_p^{\mu})], \tilde{u}(\tilde{P}_p^{\mu}) = \min[\tilde{u}^{\mu}(x_i, x_j, \theta, \vartheta)$, $(x_i, \theta), (x_j, \vartheta) \in \tilde{P}_p^{\mu}$ along the found path.

Step 11. Update the flow values in the graph $\tilde{G}_p(\tilde{\xi})$: replace the flow $\tilde{\xi}(x_j, x_i, \theta, \vartheta)$ by $\tilde{\xi}(x_j, x_i, \theta, \vartheta) - \tilde{\delta}_p^{\mu}$ along the corresponding arcs, going from (x_j, θ) to (x_i, ϑ) from $\tilde{G}_p(\tilde{\xi})$ for arcs, connecting node-time pair (x_i^{μ}, ϑ) with (x_j^{μ}, θ) in $\tilde{G}_p^{\mu}(\tilde{\xi})$, such as $((x_i^{\mu}, \vartheta), (x_j^{\mu}, \theta)) \notin \tilde{A}_p$, $((x_i^{\mu}, \vartheta), (x_j^{\mu}, \theta)) \in \tilde{A}_p^{\mu}$, and replace the flow $\tilde{\xi}(x_i, x_j, \theta, \vartheta)$ by $\tilde{\xi}(x_i, x_j, \theta, \vartheta) + \tilde{\delta}_p^{\mu}$ along the corresponding arcs, going from (x_i, θ) to (x_j, ϑ) from $\tilde{G}_p(\tilde{\xi})$ for arcs, connecting node-time pair (x_i^{μ}, θ) with (x_j^{μ}, ϑ) in $\tilde{G}_p^{\mu}(\tilde{\xi})$, such as $((x_i^{\mu}, \theta), (x_j^{\mu}, \vartheta)) \in \tilde{A}_p$, $((x_i^{\mu}, \theta), (x_j^{\mu}, \vartheta)) \in \tilde{A}_p^{\mu}$ and replace the flow value in $\tilde{G}_p$: $\tilde{\xi}(x_i, x_j, \theta, \vartheta)$ by $\tilde{\xi}(x_i, x_j, \theta, \vartheta) + \tilde{\delta}_p^{\mu} \tilde{P}_p^{\mu}$.

Step 12. If the maximum flow $\tilde{\xi}(x_i, x_j, \theta, \vartheta) + \tilde{\delta}_p^{\mu} \times \tilde{P}_p^{\mu} = \tilde{v}(p)$ from the artificial source to the artificial sink, defined by the set of paths $\tilde{P}_p^{\mu}$ in $\tilde{G}_p(\tilde{\xi})$ is found, turn to the initial dynamic graph $\tilde{G}$ as follows: reject the artificial nodes s', t' and arcs, connecting them with other nods.

Thus, the maximum flow of the value $\tilde{v}(p)$, which is equivalent to the flow from the sources (the initial node expanded to the p intervals) to the sinks (terminal node expanded to the p intervals) is obtained in $\widetilde{G}_p$ after deleting artificial nodes and each path, connecting nodes (s, ς) and $(t, \psi = \varsigma + \tau_{st}(\varsigma)), \psi \in T$, with the flow $\widetilde{\xi}(s, t, \theta, \varsigma)$ coming along, corresponds to the flow $\widetilde{\xi}_{st}(\theta)$.

Numerical example 6
Let the network, which is the part of the railway network, is represented in the form of the fuzzy directed graph, obtained from «ObjectLand» (Fig. 3.11).

It is necessary to the find the maximum flow in the dynamic graph with transit lower, upper flow bounds for 4 time periods and и represent it in the form of the fuzzy trapezoidal number, if there are basic values of arc capacities, set as fuzzy trapezoidal numbers, as shown in Fig. 2.4. Fuzzy lower and upper flow bounds and transmission times are transit and represented in Tables 3.3, 3.4 and 3.5.

Step 1. Turn to the time-expanded graph $\widetilde{G}_p$ from the graph $\widetilde{G}$ according to the *rule 3.3*, as shown in Fig. 3.12.

Step 2. Define, if the initial graph $\widetilde{G}$ has a feasible flow. Turn to the graph $\widetilde{G}_p^*$ according to the *rule 3.4*, as shown in Fig. 3.13.

Connectors, which have the same shape, link the corresponding pair of nodes in Fig. 3.13. Thus, arrow goes from each node x_5 for all time periods to the nodes x_1 at each time period, that is shown by the connectors of the same shape. Every arc, going from the nodes x_5 for all time periods to the nodes x_1 for all time periods has infinite upper flow bounds.

Step 3. Fuzzy residual network $\widetilde{G}_p^{*\mu} = (X_p^{*\mu}, \widetilde{A}_p^{*\mu})$ coincides with $\widetilde{G}_p^*$ at this step.

Step 4. Find the first augmenting path $\widetilde{P}_1^{*\mu}$ by the breadth-first-search in the residual network $\widetilde{G}_p^{*\mu}$, which coincides with $\widetilde{G}_p^*$ at the first step, presented in Fig. 3.13: $\widetilde{P}_1^{*\mu} = s^*, (x_4, 3), (x_5, 4), (x_1, 0), t^*$.

Step 5. Pass $\min([1\tilde{0}, 1\tilde{3}], [2\tilde{5}, 3\tilde{5}], \infty, [\tilde{5}, \tilde{6}]) = [\tilde{5}, \tilde{6}]$, i.e. $[\tilde{5}, \tilde{6}]$ units along the path. $\widetilde{P}_1^{*\mu} = s^*, (x_4, 3), (x_5, 4), (x_1, 0), t^*$

Fig. 3.11 Initial dynamic graph $\widetilde{G}$

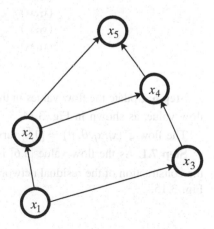

Table 3.3 Fuzzy upper flow bounds $\tilde{u}_{ij}$, depending on the flow departure time θ

Arcs of the graph	Fuzzy upper flow bounds of arcs $\tilde{u}_{ij}$ at the moment θ, units				
	0	1	2	3	4
(x_1, x_2)	$[\widetilde{10}, \widetilde{13}]$	$[\tilde{8}, \widetilde{12}]$	$[\tilde{8}, \widetilde{12}]$	$[\widetilde{10}, \widetilde{13}]$	$[\widetilde{18}, \widetilde{25}]$
(x_1, x_3)	$[\widetilde{15}, \widetilde{21}]$	$[\widetilde{15}, \widetilde{21}]$	$[\widetilde{18}, \widetilde{25}]$	$[\widetilde{18}, \widetilde{25}]$	$[\widetilde{15}, \widetilde{21}]$
(x_2, x_4)	$[\widetilde{25}, \widetilde{35}]$	$[\widetilde{20}, \widetilde{26}]$	$[\widetilde{25}, \widetilde{35}]$	$[\widetilde{18}, \widetilde{25}]$	$[\widetilde{25}, \widetilde{35}]$
(x_2, x_5)	$[\widetilde{30}, \widetilde{40}]$	$[\widetilde{25}, \widetilde{35}]$	$[\widetilde{25}, \widetilde{35}]$	$[\widetilde{30}, \widetilde{40}]$	$[\widetilde{30}, \widetilde{40}]$
(x_3, x_4)	$[\widetilde{25}, \widetilde{35}]$	$[\widetilde{25}, \widetilde{35}]$	$[\widetilde{20}, \widetilde{26}]$	$[\widetilde{20}, \widetilde{26}]$	$[\widetilde{30}, \widetilde{40}]$
(x_4, x_5)	$[\widetilde{30}, \widetilde{40}]$	$[\widetilde{25}, \widetilde{35}]$	$[\widetilde{18}, \widetilde{25}]$	$[\widetilde{25}, \widetilde{35}]$	$[\widetilde{25}, \widetilde{35}]$

Table 3.4 Fuzzy lower flow bounds $\tilde{l}_{ij}$, depending on the flow departure flow θ

Arcs of the graph	Fuzzy lower flow bounds $\tilde{l}_{ij}$ at the moment θ, units.				
	0	1	2	3	4
(x_1, x_2)	$[\tilde{5}, \tilde{6}]$	$[\tilde{0}, \tilde{0}]$	$[\tilde{0}, \tilde{0}]$	$[\tilde{5}, \tilde{6}]$	$[\tilde{0}, \tilde{0}]$
(x_1, x_3)	$[\tilde{0}, \tilde{0}]$	$[\tilde{0}, \tilde{0}]$	$[\tilde{0}, \tilde{0}]$	$[\tilde{0}, \tilde{0}]$	$[\tilde{0}, \tilde{0}]$
(x_2, x_4)	$[\tilde{0}, \tilde{0}]$	$[\tilde{0}, \tilde{0}]$	$[\tilde{0}, \tilde{0}]$	$[\tilde{8}, \widetilde{12}]$	$[\tilde{0}, \tilde{0}]$
(x_2, x_5)	$[\tilde{0}, \tilde{0}]$	$[\tilde{0}, \tilde{0}]$	$[\tilde{0}, \tilde{0}]$	$[\tilde{0}, \tilde{0}]$	$[\tilde{0}, \tilde{0}]$
(x_3, x_4)	$[\tilde{0}, \tilde{0}]$	$[\tilde{0}, \tilde{0}]$	$[\widetilde{10}, \widetilde{13}]$	$[\tilde{8}, \widetilde{12}]$	$[\tilde{0}, \tilde{0}]$
(x_4, x_5)	$[\tilde{0}, \tilde{0}]$	$[\widetilde{18}, \widetilde{25}]$	$[\tilde{0}, \tilde{0}]$	$[\tilde{0}, \tilde{0}]$	$[\tilde{0}, \tilde{0}]$

Table 3.5 Transit parameters of time τ_{ij}, depending on the flow departure time

Arcs of the graph	Transmission time along the arc of the graph τ_{ij} at the time periods θ, time units				
	0	1	2	3	4
(x_1, x_2)	1	2	3	2	2
(x_1, x_3)	5	1	3	1	1
(x_2, x_4)	6	1	3	3	3
(x_2, x_5)	2	4	3	2	2
(x_3, x_4)	5	4	1	3	1
(x_4, x_5)	5	4	1	1	2

Step 6. Update the flow values in the graph $\widetilde{G}_p^*$ and construct graph with the new flow value, as shown in Fig. 3.14.

The flow $\tilde{\xi}^*(x_i, x_j, \theta, \vartheta) = [\tilde{0}, \tilde{0}]$ turns to $\tilde{\xi}^*(x_i, x_j, \theta, \vartheta) = [\tilde{0}, \tilde{0}] + [\tilde{5}, \tilde{6}] = [\tilde{5}, \tilde{6}]$.

Step 7.1. As the flow value $[\tilde{5}, \tilde{6}]$ is less, than $\sum_{\tilde{l}(x_j, x_i, \vartheta, \theta) \neq \tilde{0}} \tilde{l}(x_i, x_j, \theta, \vartheta)$, turn to the construction of the residual network with the new flow distribution, as shown in Fig. 3.15.

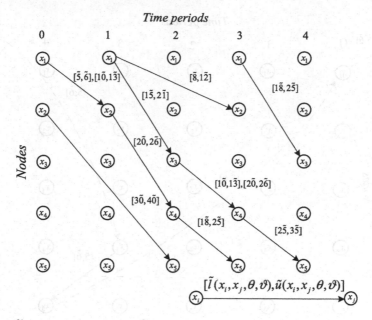

Fig. 3.12 $\widetilde{G}_p$—time-expanded graph $\widetilde{G}$

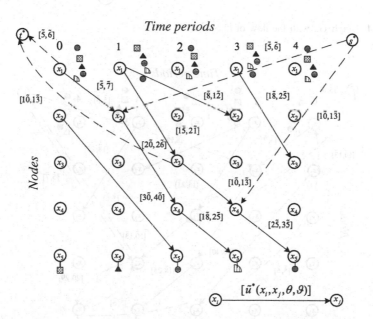

Fig. 3.13 Graph $\widetilde{G}_p^*$ without lower flow bounds with the artificial nodes and arcs

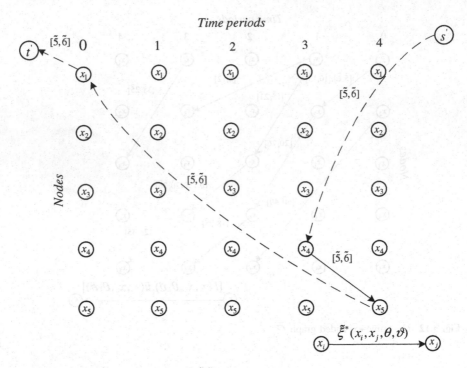

Fig. 3.14 Graph $\widetilde{G}_p^*$ with the flow of $[\tilde{5},\tilde{6}]$ units

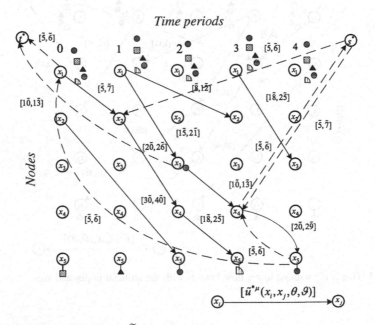

Fig. 3.15 Fuzzy residual network $\widetilde{G}_p^{*\mu}$ for the graph in Fig. 3.14

Step 4. Find the second augmenting path $\widetilde{P}_2^{*\mu}$ by the breadth-first-search in the residual network $\widetilde{G}_p^{*\mu}$, presented in Fig. 3.15: $\widetilde{P}_2^{*\mu} = s^*, (x_4, 3), (x_5, 4),$ $(x_1, 1), (x_3, 2)t^*$.

Step 5. Pass $\min([\tilde{5},\tilde{7}], [2\tilde{0},2\tilde{9}], \infty, [1\tilde{5},2\tilde{1}], [1\tilde{0},1\tilde{3}])$, i.e. $[\tilde{5},\tilde{7}]$ flow units along the path $\widetilde{P}_2^{*\mu} = s^*, (x_4, 3), (x_5, 4), (x_1, 1), (x_3, 2)t^*$.

Step 6. Update the flow values in the graph $\widetilde{G}_p^*$ and construct graph with the new flow value, as shown in Fig. 3.16.

The flow $\tilde{\xi}^*(x_i, x_j, \theta, \vartheta) = [\tilde{5},\tilde{6}]$ turns to $\tilde{\xi}^*(x_i, x_j, \theta, \vartheta) = [\tilde{5},\tilde{6}] + [\tilde{5},\tilde{7}] = [1\tilde{0},1\tilde{3}]$.

Step 7.1. As the flow value $[1\tilde{0}, 1\tilde{3}]$ is less than $\sum_{\tilde{l}(x_j, x_i, \vartheta, \theta) \neq \tilde{0}}$ $\tilde{l}(x_i, x_j, \theta, \vartheta) = [1\tilde{5}, 1\tilde{9}]$, turn to the construction of the residual network with the new flow distribution, as shown in Fig. 3.17.

Step 4. Find the third augmenting path by the breadth-first-search in the residual network $\widetilde{G}_p^{*\mu}$, presented in Fig. 3.17: $\widetilde{P}_3^{*\mu} = s^*, (x_2, 1), (x_4, 2), (x_5, 3), (x_1, 1),$ $(x_3, 2)t^*$.

Step 5. Pass $\min([\tilde{5},\tilde{6}], [2\tilde{0},2\tilde{6}], [1\tilde{8},2\tilde{5}], \infty, [1\tilde{0},1\tilde{4}], [\tilde{5},\tilde{7}]) = [\tilde{5},\tilde{6}]$, i.e. $[\tilde{5},\tilde{6}]$ flow units along the path $\widetilde{P}_3^{*\mu} = s^*, (x_2, 1), (x_4, 2), (x_5, 3), (x_1, 1), (x_3, 2)t^*$.

Step 6. Update the flow values in $\widetilde{G}_p^*$ and build a graph with the new flow value, as shown in Fig. 3.18.

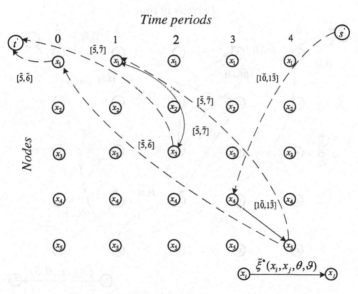

Fig. 3.16 Graph $\widetilde{G}_p^*$ with the flow of $[1\tilde{0}, 1\tilde{3}]$ units

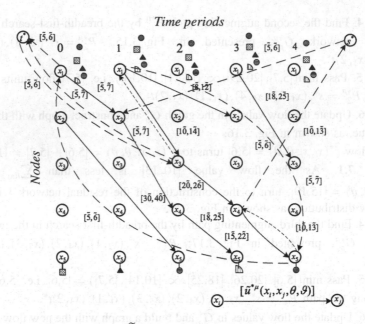

Fig. 3.17 Fuzzy residual network $\widetilde{G}_p^{*\mu}$ for the graph in Fig. 3.16

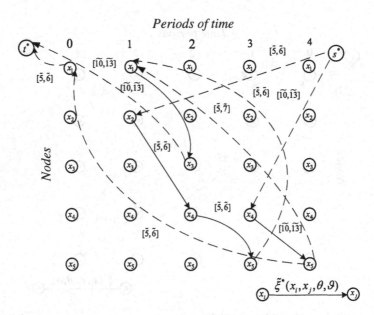

Fig. 3.18 Graph $\widetilde{G}_p^*$ with the flow of $[1\tilde{5}, 1\tilde{9}]$ units

The flow $\tilde{\xi}^*(x_i, x_j, \theta, \vartheta) = [\tilde{10}, \tilde{13}]$ turns to $\tilde{\xi}^*(x_i, x_j, \theta, \vartheta) = [\tilde{10}, \tilde{13}] + [\tilde{5}, \tilde{6}] = [\tilde{15}, \tilde{19}]$.

Step 7.2. As the flow value $[15, 19]$ is equal to $\sum_{\tilde{l}(x_j, x_i, \vartheta, \theta) \neq \tilde{0}} \tilde{l}(x_i, x_j, \theta, \vartheta) = [\tilde{15}, \tilde{19}]$, the maximum flow is found in the time-expanded graph $\widetilde{G}_p^*$, which is equal to the sum of the lower flow bounds passing along the artificial arcs. Therefore, there is a feasible flow in the time-expanded initial graph $\widetilde{G}_p$, equals the total flow, passing along the reverse arc, connecting the nodes $(x_5, \forall \theta \in T)$ with the nodes $(x_1, \forall \theta \in T)$ for all time periods, i.e. $[\tilde{5}, \tilde{6}] + [\tilde{5}, \tilde{7}] + [\tilde{5}, \tilde{6}] = [\tilde{15}, \tilde{19}]$. For the maximum flow defining in $\widetilde{G}_p$ construct the network with the flow $\widetilde{G}_p(\tilde{\xi})$, deleting artificial nodes and arcs and taking into account, that the feasible flow vector $\tilde{\xi} = (\tilde{\xi}(x_i, x_j, \theta, \vartheta))$ of the value $\tilde{\sigma}$ is defined as: $\tilde{\xi}(x_i, x_j, \theta, \vartheta) = \tilde{\xi}^*(x_i, x_j, \theta, \vartheta) + \tilde{l}(x_i, x_j, \theta, \vartheta)$, where $\tilde{\xi}^*(x_i, x_j, \theta, \vartheta)$—the flows, going along the arcs of the graph $\widetilde{G}_p^*$ after deleting the artificial nodes and arcs, connected with them, as shown in Fig. 3.19.

Step 8. Introduce artificial source and sink, connecting them by the arcs with infinite arc capacity with true sources and sinks, build fuzzy residual network $\widetilde{G}_p^\mu(\tilde{\xi})$ for the graph in Fig. 3.19, as shown in Fig. 3.20.

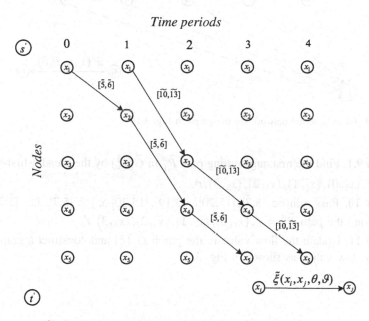

Fig. 3.19 Graph $\widetilde{G}_p(\tilde{\xi})$ with the feasible flow of $\tilde{\xi} = (\tilde{\xi}(x_i, x_j, \theta, \vartheta))$

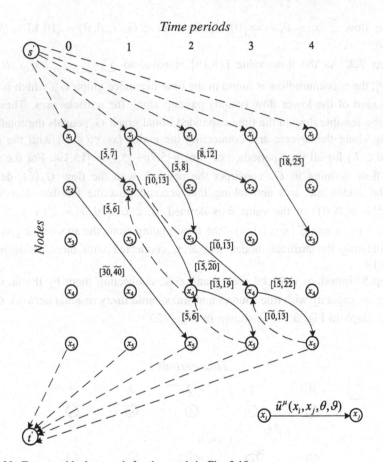

Fig. 3.20 Fuzzy residual network for the graph in Fig. 3.19

Step 9.1. Find the first augmenting path $\widetilde{P}_1^\mu$ in $\widetilde{G}_p^\mu(\tilde{\xi})$ by the breadth-first-search: $\widetilde{P}_1^\mu = s', (x_1, 0), (x_2, 1), (x_4, 2), (x_5, 3), t'$.

Step 10. Pass $\min(\infty, [\tilde{5}, \tilde{7}], [1\tilde{5}, 2\tilde{0}], [1\tilde{3}, 1\tilde{9}], [1\tilde{5}, 2\tilde{0}] \infty) = [\tilde{5}, \tilde{7}]$, i.e. $[\tilde{5}, \tilde{7}]$ flow units along the path $\widetilde{P}_1^\mu = s', (x_1, 0), (x_2, 1), (x_4, 2), (x_5, 3), t'$.

Step 11. Update the flow value in the graph $\widetilde{G}_p(\tilde{\xi})$ and construct a graph with the new flow value, as shown in Fig. 3.21.

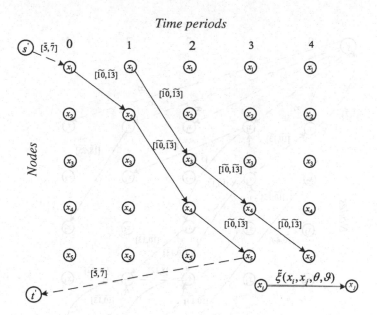

Fig. 3.21 Graph $\widetilde{G}_p(\widetilde{\xi})$ after the first augmenting path finding

Step 8. Build fuzzy residual network $\widetilde{G}_p^\mu(\widetilde{\xi})$, as shown in Fig. 3.22.

Step 9.1. Find the second augmenting path $\widetilde{P}_2^\mu$ by the breadth-first-search in $\widetilde{G}_p^\mu(\widetilde{\xi})$: $\widetilde{P}_2^\mu = s', (x_1, 1), (x_3, 2), (x_4, 3), (x_5, 4), t'$.

Step 10. Pass $\min(\infty, [\tilde{5}, \tilde{8}], [\tilde{10}, \tilde{13}], [\tilde{15}, \tilde{22}], \infty) = [\tilde{5}, \tilde{8}]$, i.e. $[\tilde{5}, \tilde{8}]$ flow units along the path $\widetilde{P}_2^\mu = s', (x_1, 1), (x_3, 2), (x_4, 3), (x_5, 4), t'$.

Step 11. Update the flow value in the graph $\widetilde{G}_p(\widetilde{\xi})$ and build a graph with the new flow value, as shown in Fig. 3.23.

Step 8. Build fuzzy residual network $\widetilde{G}_p^\mu(\widetilde{\xi})$ for the graph in Fig. 3.23, as shown in Fig. 3.24.

Step 9.2. There is no augmenting path in the fuzzy residual network in Fig. 3.24, therefore, the maximum flow is obtained in $\widetilde{G}_p(\widetilde{\xi})$, as shown in Fig. 3.25.

Turning to the dynamic graph $\widetilde{G}$ from the time-expanded static graph $\widetilde{G}_p$, we come to the conclusion. That the maximum dynamic flow for 4 time periods is equal to the flow, leaving the "node-time" pairs $(x_1, 0), (x_1, 1)$ and entering the "node-time" pairs $(x_5, 3), (x_5, 4)$, i.e. $[\tilde{25}, \tilde{34}]$ flow units, that defined by the path

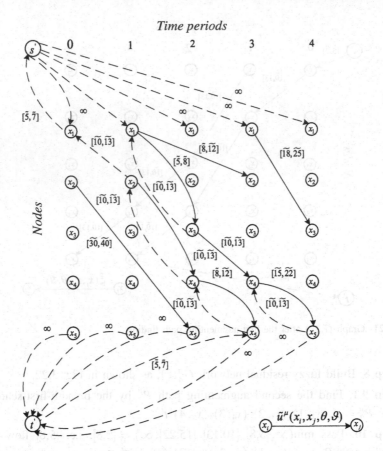

Fig. 3.22 Residual network $\widetilde{G}_p^\mu(\widetilde{\xi})$ for the graph in Fig. 3.21

$x_1 \rightarrow x_2 \rightarrow x_4 \rightarrow x_5$, with the flow starting at the time $\theta = 0$ and arriving at the sink at the time period $\theta = 3$ and the path $x_1 \rightarrow x_3 \rightarrow x_4 \rightarrow x_5$, with the flow starting at the time $\theta = 1$ and arriving at the sink at the time period $\theta = 4$.

Let us define deviation borders of the obtained fuzzy interval $[\widetilde{25}, \widetilde{34}]$, corresponding to the maximum flow in $\widetilde{G}$. The found result is between two basic neighboring values of fuzzy arc capacities: $[\widetilde{23}, \widetilde{31}]$ with the left deviation $l_1^L = 6$, right deviation—$l_1^R = 6$ and $[\widetilde{39}, \widetilde{51}]$ with the left deviation $l_2^L = 8$, the right deviation—$l_2^R = 10$. According to (2.5) we obtain:

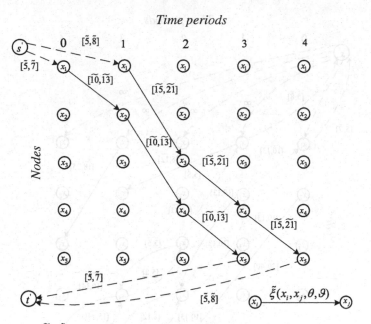

Fig. 3.23 Graph $\widetilde{G}_p(\widetilde{\xi})$ after the second augmenting path finding

$$l^L = \frac{(a_2 - a')}{(a_2 - a_1)} \times l_1^L + \left(1 - \frac{(a_2 - a')}{(a_2 - a_1)}\right) \times l_2^L$$

$$= \frac{(39 - 25)}{(39 - 23)} \times 6 + \left(1 - \frac{(39 - 25)}{(39 - 23)}\right) \times 8$$

$$= 6.25 \approx 6,$$

$$l^R = \frac{(a_2 - a')}{(a_2 - a_1)} \times l_1^R + \left(1 - \frac{(a_2 - a')}{(a_2 - a_1)}\right) \times l_2^R$$

$$= \frac{(51 - 34)}{(51 - 31)} \times 6 + \left(1 - \frac{(51 - 34)}{(51 - 31)}\right) \times 10$$

$$= 6.6 \approx 7.$$

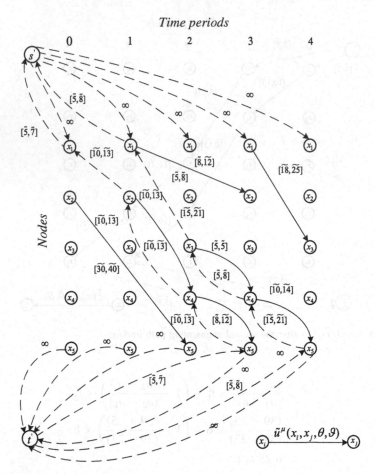

Fig. 3.24 Residual network $\widetilde{G}_p^\mu(\tilde{\xi})$ for the graph in Fig. 3.23

Therefore, the maximum flow is found in the fuzzy network, which can be represented in the form of the fuzzy trapezoidal number of (25, 34, 6, 7) units, as shown in Fig. 3.26.

According to Fig. 3.26 the maximum flow with the degree of confidence 0,7 will be within the interval [23,36], but anyway the maximum flow will be no less than 19 units and no more than 41 units.

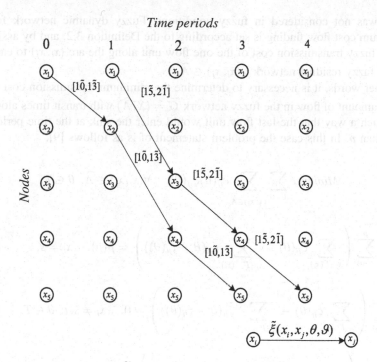

Fig. 3.25 The maximum flow in $\widetilde{G}_p$

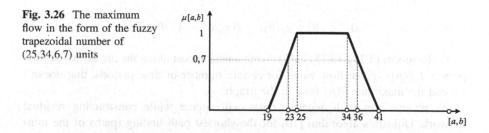

Fig. 3.26 The maximum flow in the form of the fuzzy trapezoidal number of (25,34,6,7) units

3.4 Minimum Cost Flow Finding in Dynamic Network with Fuzzy Transit Arc Capacities and Transmission Costs

Important task arising when considering dynamic networks is the task of the dynamic flow finding of the minimum cost.

Problem statement of the dynamic flow finding of the minimum cost in crisp conditions and in terms of stationary-dynamic graphs [7, 8], and the literature review of this task is considered in the first chapter of the present monograph. This task arising in fuzzy dynamic graphs, i.e. these, which parameters can vary over time, was not considered in fuzzy conditions. Fuzzy dynamic network for the minimum cost flow finding is set according to the Definition 3.2, and by assigning of the fuzzy transmission cost of the one flow unit along the arc (x_i, x_j) to each arc of the fuzzy residual network $\forall (x_i, x_j) \in \widetilde{A}$.

Other words, it is necessary to determine the minimum transmission cost of the given amount of flow in the fuzzy network $\widetilde{G} = (X, \widetilde{A})$ with transit times along the arcs such a way that the last flow unit would enter the sink at the time period not later than p. In this case the problem statement of is as follows [9]:

$$Minimize \sum_{\theta=0}^{P} \sum_{(x_i,x_j)\in\widetilde{A}} \tilde{c}_{ij}(\theta)\tilde{\xi}_{ij}(\theta), \quad \forall (x_i, x_j) \in \widetilde{A}, \ \theta \in T, \tag{3.11}$$

$$\sum_{\theta=0}^{p} \left(\sum_{x_j\in\Gamma(x_i)} \tilde{\xi}_{ij}(\theta) - \sum_{x_j\in\Gamma^{-1}(x_i)} \tilde{\xi}_{ji}(\theta - \tau_{ji}(\theta)) \right) = \tilde{\rho}(p), \quad x_i = s, \tag{3.12}$$

$$\sum_{\theta=0}^{p} \left(\sum_{x_j\in\Gamma(x_i)} \tilde{\xi}_{ij}(\theta) - \sum_{x_j\in\Gamma^{-1}(x_i)} \tilde{\xi}_{ji}(\theta - \tau_{ji}(\theta)) \right) = \tilde{0}, \quad x_i \neq s, t; \ \theta \in T, \tag{3.13}$$

$$\sum_{\theta=0}^{p} \left(\sum_{x_j\in\Gamma(x_i)} \tilde{\xi}_{ij}(\theta) - \sum_{x_j\in\Gamma^{-1}(x_i)} \tilde{\xi}_{ji}(\theta - \tau_{ji}(\theta)) \right) = -\tilde{\rho}(p), \quad x_i = t, \tag{3.14}$$

$$\tilde{0} \le \tilde{\xi}_{ij}(\theta) \le \tilde{u}_{ij}(\theta), \quad \forall (x_i, x_j) \in \widetilde{A}, \ \theta \in T. \tag{3.15}$$

In the model (3.11)–(3.15) $\tilde{c}_{ij}(\theta)$—transmission cost along the arc (x_i, x_j) at time period $\theta, \tilde{\rho}(p)$—given flow value for certain number of time periods, that doesn't exceed the maximum $\tilde{v}(p)$ flow in the graph.

As reverse arcs with negative costs will appear while constructing residual network, Dijkstra's algorithm [10] for the shortest path finding (paths of the minimum cost) can not be used. Therefore, we will use Edmonds [11] and Tomizawa's algorithm [12], transforming negative costs to nonnegative through potentials introducing, modified for the use in fuzzy dynamic graphs.

It is necessary to develop algorithm, transforming negative costs to nonnegative through potentials introducing for the minimum cost flow finding in fuzzy dynamic networks. Introduce the rule of turning to the equivalent "time-expanded" graph from the given fuzzy network and basic definitions of this method according to the *rule 3.7*, introduced in [13], and the rule of constructing the fuzzy residual network of the "time-expanded" graph (*rule 3.8*), proposed in [13] for the minimum cost flow finding.

Rule 3.7 of constructing the "time-expanded" fuzzy graph, corresponding to the *initial dynamic graph for the minimum cost flow finding with fuzzy transit arc* *capacities and transmission costs* [13].

Go to the time-expanded fuzzy static graph $\widetilde{G}_p$ from the given fuzzy dynamic graph $\widetilde{G}$ by expanding the original dynamic graph in the time dimension by making a separate copy of every node $x_i \in X$ at every time $\theta \in T$. Let $\widetilde{G}_p = (X_p, \widetilde{A}_p)$ represent fuzzy time-expanded static graph of the original dynamic fuzzy graph. The set of nodes X_p of the graph $\widetilde{G}_p$ is defined as $X_p = \{(x_i, \theta) : (x_i, \theta) \in X \times T\}$. The set of arcs $\widetilde{A}_p$ consists of arcs from each node-time pair $(x_i, \theta) \in X_p$ to every node-time pair $(x_j, \vartheta = \theta + \tau_{ij}(\theta))$ where $x_j \in \Gamma(x_i)$ and $\theta + \tau_{ij}(\theta) \leq p$. Fuzzy arc capacities $\tilde{u}(x_i, x_j, \theta, \vartheta)$ joining (x_i, θ) with (x_j, ϑ) are equal to $\tilde{u}_{ij}(\theta)$. Transmission costs $\tilde{c}(x_i, x_j, \theta, \vartheta)$ joining (x_i, θ) with (x_j, ϑ) are equal to $\tilde{c}_{ij}(\theta)$. Transit times $\tau(x_i, x_j, \theta, \vartheta)$ joining (x_i, θ) with (x_j, ϑ) are equal to $\tau_{ij}(\theta)$.

The basic notion of the method is the notion of the fuzzy modified (reduced) costs (Definition 3.4), which is synthesized definition, borrowed from [14] and modified for using in fuzzy conditions and in terms of fuzzy dynamic graphs.

Definition 3.4 *Modified (reduced) costs* $\tilde{c}^{\pi}(x_i, x_j, \theta, \vartheta)$ [14] *in the fuzzy residual network, constructed according to the "time-expanded" dynamic graph are set according to the following equation:*

$$\tilde{c}^{\pi}(x_i, x_j, \theta, \vartheta) = \tilde{c}^{\mu}(x_i, x_j, \theta, \vartheta) - \tilde{\pi}(x_i(\theta)) + \tilde{\pi}(x_j(\vartheta)), \qquad (3.16)$$

In (3.16) modified costs $\tilde{c}^{\mu}(x_i, x_j, \theta, \vartheta)$ are set according to the flow, passing along the arcs of the graph $\widetilde{G}_p$ and according the *rule 3.8.*

Rule 3.8 of constructing the fuzzy residual network in the time-expanded graph *for the minimum cost flow finding in dynamic network with fuzzy transit arc* *capacities and costs* [13].

Fuzzy residual network $\widetilde{G}_p^{\mu} = (X_p^{\mu}, \widetilde{A}_p^{\mu})$ is constructed by the "time-expanded" graph $\widetilde{G}_p = (X_p, \widetilde{A}_p)$ according to the flow values $\tilde{\xi}(x_i, x_j, \theta, \vartheta = \theta + \tau_{ij}(\theta))$ in such a way: if $\tilde{\xi}(x_i, x_j, \theta, \vartheta) < \tilde{u}^{\mu}(x_i, x_j, \theta, \vartheta)$ for the arcs going from the nodes (x_i, θ) to (x_j, ϑ) in $\widetilde{G}_p$, then include the arc, connecting the nodes (x_i^{μ}, θ) to the nodes (x_j^{μ}, ϑ) in $\widetilde{G}_p^{\mu}$ with residual fuzzy arc capacity $\tilde{u}^{\mu}(x_i, x_j, \theta, \vartheta) = \tilde{u}(x_i, x_j, \theta, \vartheta) - \tilde{\xi}(x_i, x_j, \theta, \vartheta)$, the transit time $\tau^{\mu}(x_i, x_j, \vartheta, \theta) = -\tau(x_i, x_j, \theta, \vartheta)$ and modified cost $\tilde{c}^{\pi}(x_i, x_j, \theta, \vartheta) = \tilde{c}^{\mu}(x_i, x_j, \theta, \vartheta) - \tilde{\pi}(x_i(\theta)) + \tilde{\pi}(x_j(\vartheta))$, where $\tilde{c}^{\mu}(x_i, x_j, \theta, \vartheta) = \tilde{c}(x_i, x_j, \theta, \vartheta)$. If $\tilde{\xi}(x_i, x_j, \theta, \vartheta) > \tilde{0}$ for the arcs, connecting the nodes (x_i, θ) and (x_j, ϑ) in $\widetilde{G}_p$, then include the arc, connecting the nodes (x_j^{μ}, ϑ) with (x_i^{μ}, θ) in $\widetilde{G}_p^{\mu}$ with residual fuzzy arc capacity $\tilde{u}^{\mu}(x_j, x_i, \vartheta, \theta) = \tilde{\xi}(x_i, x_j, \theta, \vartheta)$, transit time $\tau^{\mu}(x_j, x_i, \vartheta, \theta) = -\tau(x_i, x_j, \theta, \vartheta)$ and modified cost $\tilde{c}^{\pi}(x_j, x_i, \vartheta, \theta) = \tilde{c}^{\mu}(x_j, x_i, \vartheta, \theta) + \tilde{\pi}(x_i(\theta)) - \tilde{\pi}(x_j(\vartheta))$, where $\tilde{c}^{\mu}(x_j, x_i, \vartheta, \theta) = -\tilde{c}(x_i, x_j, \theta, \vartheta)$.

For some cycle $\widetilde{H}_p^\mu$ in $\widetilde{G}_p^\mu$ is true:

$$\sum_{(x_i,x_j,\theta,\vartheta)\in\widetilde{H}_p^\mu} \tilde{c}^\pi(x_i,x_j,\theta,\vartheta) = \sum_{(x_i,x_j,\theta,\vartheta)\in\widetilde{H}_p^\mu} \tilde{c}^\mu(x_i,x_j,\theta,\vartheta).$$

Let us consider Property 3.1, synthesized through the property, presented in [14] and modified for the use in fuzzy conditions and in terms of dynamic networks.

Property 3.1 *For the path from s' to t' in $\widetilde{G}_p^\mu$ the equality is valid:*

$$\sum_{(x_i,x_j,\theta,\vartheta)\in\widetilde{P}_p^\mu} \tilde{c}^\pi(x_i,x_j,\theta,\vartheta) = \tilde{\pi}(t'+\varepsilon+\tau_{s't'}(\varepsilon)) - \tilde{\pi}(s',\varepsilon)$$

$$+ \sum_{(x_i,x_j,\theta,\vartheta)\in\widetilde{P}_p^\mu} \tilde{c}^\mu(x_i,x_j,\theta,\vartheta). \tag{3.17}$$

Node potentials don't vary the minimum cost path among any pair of nodes according to (3.17).

For the proof of the flow optimality, obtained after potentials introducing give the Property 3.2, which is the synthesis of the property, presented in [14] and modified for the use in fuzzy dynamic networks.

Property 3.2 *The flow $\tilde{\xi}^*(x_i,x_j,\theta,\vartheta)$ is optimal if and only if, when the potentials vector exists, such as $\tilde{c}^\pi(x_i,x_j,\theta,\vartheta)\geq\tilde{0}$ for all $\forall(x_i^\mu,x_j^\mu,\theta,\vartheta)\in\widetilde{A}_p^\mu$.*

Proof Let us assume that there are such flow values $\tilde{\pi}$, that $\tilde{c}^\pi(x_i,x_j,\theta,\vartheta)\geq\tilde{0}$ for all $\forall(x_i^\mu,x_j^\mu,\theta,\vartheta)\in\widetilde{A}_p^\mu$. Let $\widetilde{H}_p^\mu$ be the cycle in $\widetilde{G}_p^\mu$, then:

$$\tilde{c}\left(\widetilde{H}_p^\mu\right) = \sum_{(x_i,x_j,\theta,\vartheta)\in\widetilde{H}_p^\mu} \tilde{c}^\mu(x_i,x_j,\theta,\vartheta)$$

$$= \sum_{(x_i,x_j,\theta,\vartheta)\in\widetilde{H}_p^\mu} \tilde{c}^\pi(x_i,x_j,\theta,\vartheta)\geq\tilde{0}.$$

Therefore, there is no cycle of the negative cost in $\widetilde{G}_p^\mu$ and consequently the flow is optimal. The corollary is proved.

Let us show that if there is no cycle of the negative cost in the residual network $\widetilde{G}^\mu$, then there is a flow vector with non-negative reduced costs.

Assume all nodes in the residual network $\widetilde{G}_p^\mu$ can be reachable from (s',ς). Let $\tilde{\gamma}^\mu(s',x_j,\varsigma,\theta=\varsigma+\tau_{s'x_j}(\varsigma))$ determines the length of the shortest path from (s',ς) to (x_j^μ,θ) in $\widetilde{G}_p^\mu$. If $\widetilde{G}_p^\mu$ has no cycle of the negative cost, then $\tilde{\gamma}^\mu(s',x_j,\varsigma,\theta)\leq \tilde{\gamma}^\mu(s',x_i,\varsigma,\theta)+\tilde{c}^\mu(x_i,x_j,\theta,\vartheta)$. Therefore, setting $\tilde{\pi}(x_j,\theta)=-\tilde{\gamma}^\mu(s',x_j,\varsigma,\theta)$ we obtain:

$$\tilde{c}^{\pi}(x_i, x_j, \theta, \vartheta) = \tilde{c}^{\mu}(x_i, x_j, \theta, \vartheta) - \tilde{\pi}(x_i, \theta) + \tilde{\pi}(x_j, \theta)$$
$$= \tilde{c}^{\mu}(x_i, x_j, \theta, \vartheta) + \tilde{\gamma}^{\mu}(s', x_i, \varsigma, \theta) - \tilde{\gamma}^{\mu}(s', x_j, \varsigma, \theta) \geq \tilde{0}.$$

For the flow $\tilde{\xi}(x_i, x_j, \theta, \vartheta)$ and node $(x_i, \theta) \in X$ the node condition (x_i, θ) is set as:

$$\tilde{e}(x_i, \theta) = \tilde{\rho}(x_i, \theta) + \sum_{(x_i, x_j, \theta, \vartheta) \in \tilde{A}_p} \tilde{\xi}(x_i, x_j, \theta, \vartheta) - \sum_{(x_j, x_i, \vartheta, \theta) \in \tilde{A}_p} \tilde{\xi}(x_j, x_i, \vartheta, \theta). \quad (3.18)$$

According to (3.18):

If $\tilde{e}(x_i, \theta) > \tilde{0}$, then the condition in the node is excess.

If $\tilde{e}(x_i, \theta) < \tilde{0}$, then the condition in the node is deficit.

If $\tilde{e}(x_i, \theta) = \tilde{0}$, then the condition in the node is balance.

For further proof of the equality $\tilde{\pi}(x_j, \theta) = -\tilde{\gamma}^{\mu}(s', x_j, \varsigma, \theta)$ give Lemmas 3.1 and 3.2, which are modifications of the lemmas given in [14] for using in fuzzy conditions and in terms of fuzzy dynamic networks.

Lemma 3.1 *Let us assume that the flow $\tilde{\xi}(x_i, x_j, \theta, \vartheta)$ and potentials, assigned to the nodes, satisfy the optimality criteria of reduced costs: $\tilde{c}^{\pi}(x_i, x_j, \theta, \vartheta) = \tilde{c}^{\mu}(x_i, x_j, \theta, \vartheta) - \tilde{\pi}(x_i, \theta) + \tilde{\pi}(x_j, \vartheta) \geq \tilde{0}$ for $\forall(x_i^{\mu}, x_j^{\mu}, \theta, \vartheta) \in \tilde{A}_p^{\mu}$. For each node $(x_j^{\mu}, \vartheta) \in X^{\mu}$ let $\tilde{\gamma}^{\mu}(s', x_j, \varsigma, \theta)$ define the length of shortest path from (s', ς) to (x_j^{μ}, ϑ) in $\tilde{G}_p^{\mu}$, assuming $\tilde{c}^{\pi}(x_i, x_j, \theta, \vartheta)$ as arc lengths. The following properties follow from this statement.*

The flow $\tilde{\xi}(x_i, x_j, \theta, \vartheta)$ satisfies the optimality criteria of the reduced costs relative to the potentials, i.e. the following condition is satisfied:

$$\tilde{\pi}'(x_j, \vartheta) = \tilde{\pi}(x_j, \vartheta) - \tilde{\gamma}^{\mu}(s', x_j, \varsigma, \theta) \quad \text{for} \quad \forall(x_j, \vartheta) \in X. \quad (3.19)$$

Proof As $\tilde{\gamma}^{\mu}(s', x_j, \varsigma, \theta)$ are lengths of the shortest paths from (s', ς) to (x_j^{μ}, ϑ) in $\tilde{G}_p^{\mu}$ in (3.19) for $\forall(x_i, x_j, \theta, \vartheta) \in \tilde{A}_p^{\mu}$, then the following condition is satisfied:

$$\tilde{\gamma}^{\mu}(s', x_j, \varsigma, \theta) \leq \tilde{\gamma}^{\mu}(s', x_i, \varsigma, \theta) + \tilde{c}^{\pi}(x_i, x_j, \theta, \vartheta) \quad (3.20)$$

Therefore, based on (3.20) it is true:

$$\tilde{c}^{\pi'}(x_i, x_j, \theta, \vartheta) = \tilde{c}^{\mu}(x_i, x_j, \theta, \vartheta) - \tilde{\pi}'(x_i, \theta) + \tilde{\pi}'(x_j, \vartheta)$$
$$= \tilde{c}^{\mu}(x_i, x_j, \theta, \vartheta) - (\tilde{\pi}(x_i, \theta)) - \tilde{\gamma}^{\mu}(s', x_i, \varsigma, \theta)) + (\tilde{\pi}(x_j, \vartheta)) - \tilde{\gamma}^{\mu}(s', x_i, \varsigma, \theta))$$
$$= \tilde{c}^{\pi}(x_i, x_j, \theta, \vartheta) + \tilde{\gamma}^{\mu}(s', x_i, \varsigma, \theta) - \tilde{\gamma}^{\mu}(s', x_j, \varsigma, \theta) \geq \tilde{0}.$$

Reduced costs $\tilde{c}^{\pi'}(x_i, x_j, \theta, \vartheta)$ are equal to $\tilde{0}$ for $\forall (x_i, x_j, \theta, \vartheta) \in \widetilde{A}_p^\mu$ in the shortest path from s' to t'.

Proof Let $(x_i^\mu, x_j^\mu, \theta, \vartheta)$ be the shortest path from s' to t'. Then

$$
\begin{aligned}
\tilde{c}^{\pi'}(x_i, x_j, \theta, \vartheta) &= \tilde{c}^\mu(x_i, x_j, \theta, \vartheta) - (\tilde{\pi}(x_i, \theta) - \tilde{\gamma}^\mu(s', x_i, \varsigma, \theta)) \\
&+ (\tilde{\pi}(x_j, \vartheta) - \tilde{\gamma}^\mu(s', x_j, \varsigma, \vartheta)) = \tilde{c}^\pi(x_i, x_j, \theta, \vartheta) \\
&+ \tilde{\gamma}^\mu(s', x_i, \varsigma, \theta) - \tilde{\gamma}^\mu(s', x_j, \varsigma, \vartheta) = \tilde{0},
\end{aligned}
$$

as $\tilde{\gamma}^\mu(s', x_j, \varsigma, \theta) = \tilde{\gamma}^\mu(s', x_i, \varsigma, \theta) + \tilde{c}^\pi(x_i, x_j, \theta, \vartheta)$, i.e. the arc $(x_i, x_j, \theta, \vartheta) \in \widetilde{A}_p^\mu$ is the arc of the shortest path. Lemma is proved.

Based on the formulated lemmas and properties represent algorithm of solving this task.

Algorithm of the minimum cost flow finding in dynamic network with transit fuzzy arc capacities and transmission costs

Consider algorithm, realized minimum cost flow finding in dynamic network with transit fuzzy arc capacities and transmission costs in terms of partial uncertainty [13].

Step 1. Set initial values of flows, potentials and node conditions $\tilde{\xi}_{ij}(\theta) = \tilde{0}$, $\tilde{\pi}(x_i) = \tilde{0}$, $\tilde{e}(x_i) = \tilde{\rho}(x_i)$ for $\forall x_i \in X$, $(x_i, x_j) \in \widetilde{A}$. Consider, that for intermediate nodes $\tilde{\rho}(x_i) = \tilde{0}$, $\tilde{\rho}(x_i) > \tilde{0}$ for the source, $\tilde{\rho}(x_i) < \tilde{0}$ for the sink.

Step 2. Turn to the "time-expanded" on p intervals fuzzy static graph $\widetilde{G}_p$ from the given fuzzy dynamic graph $\widetilde{G} = (X, \widetilde{A})$ by expanding initial dynamic graph for given number of time periods by making separate copy of each node $x_i \in X$ at each time period $\theta \in T$ according to the *rule 3.7*.

It is necessary to find the maximum flow of the minimum cost, leaving the group of sources at all time periods and entering the sinks at all time periods not later than p. Introduce artificial source s' and sink and connect s' by arcs with each true source s' and t' with each true sink. Artificial arcs, going from artificial nodes have infinite arc capacity.

Step 3. Set initial values of flows, potentials and node conditions $\tilde{\xi}(x_i, x_j, \theta, \vartheta) = \tilde{0}$, $\tilde{\pi}(x_i(\theta)) = \tilde{0}$, $\tilde{e}(x_i(\theta)) = \tilde{\rho}_i$ for $\forall (x_i, \theta) \in X_p$, $(x_i, x_j, \theta, \vartheta) \in \widetilde{A}_p$ in the "time-expanded" graph $\widetilde{G}_p$. Note, that true sources, extended for the given number of time periods will have original exceeds equal $\tilde{0}$, i.e. these nodes are intermediate nodes. Initial values of exceeds in s' and deficits in t' in $\widetilde{G}_p$ are equivalent to initial values of exceeds in s and deficit in t in $\widetilde{G}$.

Step 4. Check the existence of the exceed in the node s'.

4.1. If there is the exceed in the node s', i.e. $\tilde{e}(s') > \tilde{0}$, turn to the step 5.

4.2. If there is no exceed in the node s', therefore, given flow value is found, turn to the step 11.

Step 5. Construct fuzzy residual network $\widetilde{G}_p^\mu$ for the "time-expanded" graph $\widetilde{G}_p$ according to the flow values, passing along the arcs of the graph due to the *rule 3.8*. In the first step fuzzy residual network coincides with the original one due to the equality of flows, and reduced costs coincide with initial ones, since $\tilde{\pi}(x_i(\theta)) = \tilde{0}$.

Step 6. Determine the minimum cost path from s' to t' using E. Dijkstra's algorithm in the residual network, based on the reduced costs $\tilde{c}^\pi(x_i, x_j, \theta, \vartheta)$.

6.1. If the path exists, i.e. the permanent label is assigned to the node t', stop and go to the step 6.

6.2. If the path does not exist, i.e. vertex t' is not reachable, then the task has no solution, exit.

Step 7. Let t' be the node with deficit, which has constant label. Obtain the path $\widetilde{P}^\mu(s, t, \omega, \omega + \tau_{st}(\omega))$ from s to t.

Step 8. Define new node potentials as:

$$\tilde{\pi}(x_i, \theta) = \begin{cases} \tilde{\pi}(x_i, \theta) - \tilde{\gamma}^\mu(s', x_i, \varsigma, \theta), \textit{if the node} \\ x_i \textit{ has permanent label,} \\ \tilde{\pi}(x_i, \theta) - \tilde{\gamma}^\mu(s', t', \varsigma, \psi = \varsigma + \tau_{s't'}), \psi \in T, \textit{in other case.} \end{cases} \quad (3.21)$$

Step 9. Push $\tilde{\delta}_p^\mu = \min\{\tilde{u}(\widetilde{P}_p^\mu), \tilde{e}(s'), -\tilde{e}(t')\}$ flow units along the path $\widetilde{P}^\mu$, where $\tilde{u}(\widetilde{P}^\mu)$ is the arc capacity of the path, defining as minimum of the arc capacities of this path: $\tilde{u}(\widetilde{P}_p^\mu) = \min[\tilde{u}(x_i, x_j, \theta, \vartheta)], (x_i, x_j, \theta, \vartheta) \in \widetilde{P}_p^\mu$.

Step 10. Update flow values in the graph $\widetilde{G}_p$: for arcs, connecting the node-time pairs (x_i^μ, ϑ) with (x_j^μ, θ), such as $(x_i^\mu, x_j^\mu) \notin \widetilde{A}, (x_i^\mu, x_j^\mu) \in \widetilde{A}^\mu$ in $\widetilde{G}_p^\mu$ change the flow $\tilde{\xi}(x_j, x_i, \theta, \vartheta)$ along the corresponding arcs from (x_j, θ) to (x_i, ϑ) of $\widetilde{G}_p$ from $\tilde{\xi}(x_j, x_i, \theta, \vartheta)$ to $\tilde{\xi}(x_j, x_i, \theta, \vartheta) - \tilde{\delta}_p^\mu$. For the arcs, connecting the node-time pair (x_i^μ, θ) with (x_j^μ, ϑ), such as $(x_i^\mu, x_j^\mu) \in \widetilde{A}, (x_i^\mu, x_j^\mu) \in \widetilde{A}^\mu$ in $\widetilde{G}_p^\mu$ change the flow $\tilde{\xi}(x_i, x_j, \theta, \vartheta)$ along the corresponding arcs from (x_i, θ) to (x_j, ϑ) of $\widetilde{G}_p$ from $\tilde{\xi}(x_i, x_j, \theta, \vartheta)$ to $\tilde{\xi}(x_i, x_j, \theta, \vartheta) + \tilde{\delta}_p^\mu$, replace the flow value in the graph $\widetilde{G}_p$ $\tilde{\xi}(x_i, x_j, \theta, \vartheta) \rightarrow \tilde{\xi}(x_i, x_j, \theta, \vartheta) + \tilde{\delta}_p^\mu \times \widetilde{P}_p^\mu$ and turn to the step 4.

Step 11. If the flow value $\tilde{\xi}(x_i, x_j, \theta, \vartheta) + \tilde{\delta}_p^\mu \times \widetilde{P}_p^\mu = \tilde{\rho}(p)$ in $\widetilde{G}_p$ is found in $\widetilde{G}_p$, defined by the set of the paths $\widetilde{P}_p^\mu$ from the artificial source to the artificial sink, replace modified costs $\tilde{c}^\pi(x_i, x_j, \theta, \vartheta)$ by initial ones $\tilde{c}(x_i, x_j, \theta, \vartheta)$ and turn to the initial dynamic graph $\widetilde{G}$ as follows: delete artificial nodes with potentials and arcs, connecting them with other nodes. Thus, given flow value from the source to the sink is obtained in the initial graph $\widetilde{G}$ and each path, connecting nodes (s, θ) and $(t, \varsigma = \theta + \tau_{st}(\theta)), \varsigma \in T$, with the flow $\tilde{\xi}(s, t, \theta, \varsigma)$ of the cost $\tilde{\xi}_{st}(\theta)$ coming along corresponds to the flow $\tilde{\xi}_{st}(\theta)$ of the cost $\tilde{c}(\tilde{\xi}_{st}(\theta))$. Find its minimum cost $\tilde{c}(\tilde{\xi}(x_i, x_j, \theta, \vartheta) + \tilde{\delta}_p^\mu \times \widetilde{P}_p^\mu)$.

Since using the algorithm of E. Dijkstra we are interested in the way of the minimum cost from the source to the sink, there is no need to find the minimum cost path from the source to all vertices (as it is traditionally applied in the algorithm of E. Dijkstra). Consequently, the termination criterion of the minimum cost path algorithm of E. Dijkstra is not attributing the permanent marks to all vertices, and the attribution of the permanent mark to the sink node. Then updating the node potentials, as it is presented in the step 6 of the minimum cost flow algorithm will occur according to (3.21).

Consider example, describing the presented algorithm.

Numerical example 7

Let the network be the part of the railway network and presented in the form of the fuzzy directed graph, obtained from GIS «ObjectLand» (Fig. 3.11). The node x_1 is the source, x_5 is the sink.

Values of transit fuzzy arc capacities, transmission costs and crisp transit times are assigned to the arcs of the graph in Tables 3.6, 3.7 and 3.8. It is necessary to define the minimum transportation cost of the fuzzy flow $\tilde{\rho}(p) = 2\tilde{0}$ units for 3 time periods and represent the result in the form of the fuzzy triangular number, if there are basic values of arc capacities and transmission costs, represented in Figs. 2.27–2.30.

Step 2. Turn to the "time-expanded" static graph of the initial dynamic graph according to the *rule 3.7* and set initial zero values of flows, potentials and exceeds $\tilde{\xi}_{ij} = \tilde{0}, \pi(x_i) = \tilde{0}, e(x_i) = \tilde{\rho}(x_i)$ for $\forall x_i \in X, (x_i, x_j) \in \widetilde{A}$, as shown in Fig. 3.27.

Step 4.1. Determine the condition of the node s': $e(s') = 2\tilde{0} + \tilde{0} - \tilde{0} = 2\tilde{0}$, i.e. the node s': has an exceed, therefore, turn to the step 5.

Step 5. Fuzzy residual network $\widetilde{G}_p^\mu$ coincides with the initial one, presented in Fig. 3.27, since the equality of arc flows $\tilde{0}$.

Step 6. Determine the minimum cost path from s' to t' via E. Dijkstra's algorithm in residual network, presented in Fig. 3.27.

Let $\tilde{l}(x_i, \theta) = \tilde{0}$—the label of the node x_i at the time θ.

Step 1. Set $\tilde{l}(s') = \tilde{0}$ and consider this label as constant one. Let $\tilde{l}(x_i, \forall \theta \in T) = \infty, \forall(x_i, \theta \in T) \neq s'$ and consider these labels as temporary. Let $p = s'$.

Table 3.6 Fuzzy arc capacities $\tilde{u}_{ij}$, depend on the flow departure time θ	Arcs of the graph	Fuzzy arc capacities $\tilde{u}_{ij}$ at the moment θ, units.			
		0	1	2	3
	(x_1, x_2)	$\widetilde{10}$	$\widetilde{20}$	$\tilde{8}$	$\widetilde{10}$
	(x_1, x_3)	$\widetilde{18}$	$\widetilde{15}$	$\widetilde{18}$	$\widetilde{18}$
	(x_2, x_4)	$\widetilde{25}$	$\widetilde{20}$	$\tilde{5}$	$\widetilde{18}$
	(x_2, x_5)	$\widetilde{30}$	$\widetilde{25}$	$\widetilde{25}$	$\widetilde{30}$
	(x_3, x_4)	$\widetilde{25}$	$\widetilde{25}$	$\widetilde{20}$	$\widetilde{20}$
	(x_4, x_5)	$\widetilde{30}$	$\widetilde{25}$	$\widetilde{30}$	$\widetilde{25}$

Table 3.7 Fuzzy transmission costs $\tilde{c}_{ij}$, depend on the flow departure time θ

Arcs of the graph	Fuzzy transmission costs $\tilde{c}_{ij}$ at the moment θ, conventional units.			
	0	1	2	3
(x_1, x_2)	$\widetilde{40}$	$\widetilde{30}$	$\widetilde{60}$	$\widetilde{30}$
(x_1, x_3)	$\widetilde{50}$	$\widetilde{30}$	$\widetilde{40}$	$\widetilde{70}$
(x_2, x_4)	$\widetilde{60}$	$\widetilde{40}$	$\widetilde{50}$	$\widetilde{60}$
(x_2, x_5)	$\widetilde{70}$	$\widetilde{55}$	$\widetilde{75}$	$\widetilde{60}$
(x_3, x_4)	$\widetilde{50}$	$\widetilde{35}$	$\widetilde{50}$	$\widetilde{50}$
(x_4, x_5)	$\widetilde{80}$	$\widetilde{25}$	$\widetilde{30}$	$\widetilde{60}$

Table 3.8 Transit times τ_{ij}, depend on the flow departure time θ

Arcs of the graph	Transit times τ_{ij}, depend on the flow departure time θ, time units.			
	0	1	2	3
(x_1, x_2)	1	1	3	2
(x_1, x_3)	4	3	1	1
(x_2, x_4)	4	1	3	3
(x_2, x_5)	2	4	1	2
(x_3, x_4)	5	4	3	3
(x_4, x_5)	1	4	1	1

First iteration

Step 2. Change the labels for all nodes with temporary labels according to the following equality:

$$\tilde{l}(x_i, \theta) = \min[\tilde{l}(x_i, \theta), \tilde{l}(p) + \tilde{c}(p, (x_i, \theta))] \tag{3.22}$$

$\Gamma(p) = \Gamma(s') = \{(x_1, 0), (x_1, 1), (x_1, 2), (x_1, 3)\}$—all labels are temporary. We obtain the following equation from (3.22):

$$\tilde{l}(x_1, 0) = \min[\infty, \tilde{0}^+, +\tilde{0}] = \tilde{0},$$
$$\tilde{l}(x_1, 1) = \min[\infty, \tilde{0}^+, +\tilde{0}] = \tilde{0},$$
$$\tilde{l}(x_1, 2) = \min[\infty, \tilde{0}^+, +\tilde{0}] = \tilde{0},$$
$$\tilde{l}(x_1, 3) = \min[\infty, \tilde{0}^+, +\tilde{0}] = \tilde{0}.$$

Step 3. Define the minimum value for nodes

$(x_1, 0), (x_1, 1), (x_1, 2), (x_1, 3), (x_2, 0), (x_2, 1), (x_2, 2), (x_2, 3), (x_3, 0), (x_3, 1), (x_3, 2),$
$(x_3, 3), (x_4, 0), (x_4, 1), (x_4, 2), (x_4, 3), (x_5, 0), (x_5, 1), (x_5, 2), (x_5, 3), t',$

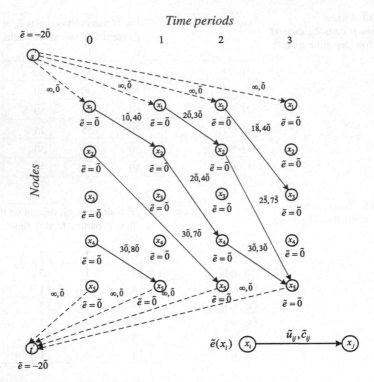

Fig. 3.27 «Time-expanded» graph $\widetilde{G}_p$ with the given node balances

as $\min[\tilde{0}, \tilde{0}, \tilde{0}, \tilde{0}, \infty, \infty, \infty, \infty, \infty, \infty, \infty, \infty, \infty, \infty, \infty, \infty, \infty, \infty, \infty, \infty, \infty] = \tilde{0}$.
Let the minimal label will be $(x_1, 0)$.

Step 4. Consider the label of the node $(x_1, 0)$ as constant, $\tilde{l}(x_1, 0) = \tilde{0}^+, p = (x_1, 0)$.
Step 5. $p \neq t'$, turn to the step 2.

Second iteration

Step 2. Change the labels for all nodes with temporary labels.
$\Gamma(p) = \Gamma(x_1, 0) = \{(x_2, 1)\}$—the label is temporary. We obtain from (3.22):

$$\tilde{l}(x_1, 0) = \min[\infty, \tilde{0}^+ + \tilde{0}] = \tilde{0},$$
$$\tilde{l}(x_2, 1) = \min[\infty, \tilde{0} + 4\tilde{0}] = 4\tilde{0}.$$

Step 3. Define the minimum value for nodes:

$(x_1, 1), (x_1, 2), (x_1, 3), (x_2, 0), (x_2, 1), (x_2, 2), (x_2, 3), (x_3, 0), (x_3, 1).(x_3, 2), (x_3, 3),$
$(x_4, 0), (x_4, 1), (x_4, 2), (x_4, 3), (x_5, 0), (x_5, 1), (x_5, 2), (x_5, 3), t',$

as $\min[\tilde{0}, \tilde{0}, \tilde{0}, \infty, 4\tilde{0}, \infty, \infty, \infty, \infty, \infty, \infty, \infty, \infty, \infty, \infty, \infty, \infty, \infty, \infty, \infty] = \tilde{0}.$
Choose the node with the label $(x_1, 1)$ as minimal.

Step 4. Consider this label $(x_1, 1)$ as constant, $\tilde{l}(x_1, 1) = \tilde{0}^+, \tilde{p} = (x_1, 1).$
Step 5. $p \neq t'$, turn to the step 2.

Third iteration

Step 2. Change the labels for all nodes with temporary labels.
$\Gamma(p) = \Gamma(x_1, 1) = \{(x_2, 2)\}$—the label is temporary. We obtain from (3.22):

$$\tilde{l}(x_2, 2) = \min[\infty, \tilde{0} + 3\tilde{0}] = 3\tilde{0}.$$

Step 3. Define the minimum value for nodes:

$(x_1, 2), (x_1, 3), (x_2, 0), (x_2, 1), (x_2, 2), (x_2, 3), (x_3, 0), (x_3, 1).(x_3, 2), (x_3, 3), (x_4, 0),$
$(x_4, 1), (x_4, 2), (x_4, 3), (x_5, 0), (x_5, 1), (x_5, 2), (x_5, 3), t'$

as $\min[\tilde{0}, \tilde{0}, \infty, 4\tilde{0}, 3\tilde{0}, \infty, \infty, \infty, \infty, \infty, \infty, \infty, \infty, \infty, \infty, \infty, \infty, \infty, \infty] = \tilde{0}.$
Choose the node with the label $(x_1, 2)$ as minimal.

Step 4. Consider the label $(x_1, 2)$ as constant one, $\tilde{l}(x_1, 2) = \tilde{0}^+, \tilde{p} = (x_1, 2).$
Step 5. $p \neq t'$, turn to the step 2.

Fourth iteration

Step 2. Change the labels for all nodes with temporary labels.
$\Gamma(p) = \Gamma(x_1, 2) = \{(x_3, 3)\}$—the label is temporary. We obtain from (3.22):

$$\tilde{l}(x_3, 3) = \min[\infty, \tilde{0} + 4\tilde{0}] = 4\tilde{0}.$$

Step 3. Define the minimum value for nodes

$(x_1, 3), (x_2, 0), (x_2, 1), (x_2, 2), (x_2, 3), (x_3, 0), (x_3, 1).(x_3, 2), (x_3, 3), (x_4, 0), (x_4, 1),$
$(x_4, 2), (x_4, 3), (x_5, 0), (x_5, 1), (x_5, 2), (x_5, 3), t',$

as $\min[\tilde{0}, \infty, 4\tilde{0}, 3\tilde{0}, \infty, \infty, \infty, \infty, 4\tilde{0}, \infty, \infty, \infty, \infty, \infty, \infty, \infty, \infty, \infty] = \tilde{0}.$ The
label $(x_1, 3)$ has the minimum value.

Step 4. Consider the node label $(x_1, 3)$ as constant, $\tilde{l}(x_1, 3) = \tilde{0}^+, \tilde{p} = (x_1, 3).$
Step 5. $p \neq t'$, turn to the *step 2*.

Fifth iteration

Step 2. Change the labels for all nodes with temporary labels.

$$\Gamma(p) = \Gamma(x_1, 2) = \{\varnothing\}$$

Step 3. Define the minimum value for nodes

$$(x_2, 0), (x_2, 1), (x_2, 2), (x_2, 3), (x_3, 0), (x_3, 1), (x_3, 2), (x_3, 3), (x_4, 0), (x_4, 1), (x_4, 2), (x_4, 3),$$
$$(x_5, 0), (x_5, 1), (x_5, 2), (x_5, 3), t'$$

as $\min[\infty, 4\tilde{0}, 3\tilde{0}, \infty, \infty, \infty, \infty, 4\tilde{0}, \infty, \infty, \infty, \infty, \infty, \infty, \infty, \infty, \infty] = 3\tilde{0}$. The label $(x_2, 2)$ has the minimum value.

Step 4. Consider the node label $(x_2, 2)$ as constant, $\tilde{l}(x_2, 2) = 3\tilde{0}$, $\tilde{p} = (x_2, 2)$.

Step 5. $p \neq t'$, turn to the *step 2*.

Sixth iteration

Step 2. Change the labels for all nodes with temporary labels
$\Gamma(p) = \Gamma(x_2, 2) = \{(x_5, 3)\}$—the node $(x_5, 3)$ has temporary label. We obtain from (3.22):

$$\tilde{l}(x_5, 3) = \min[\infty, 3\tilde{0} + 7\tilde{5}] = 10\tilde{5}.$$

Step 3. Define the minimum value for nodes

$$(x_2, 0), (x_2, 1), (x_2, 3), (x_3, 0), (x_3, 1).(x_3, 2), (x_3, 3), (x_4, 0), (x_4, 1), (x_4, 2), (x_4, 3),$$
$$(x_5, 0), (x_5, 1), (x_5, 2), (x_5, 3), t'$$

as $\min[\infty, 4\tilde{0}, \infty, \infty, \infty, \infty, 4\tilde{0}, \infty, \infty, \infty, \infty, \infty, \infty, \infty, 10\tilde{5}, \infty] = 4\tilde{0}$. The label $(x_2, 1)$ has the minimum value.

Step 4. Consider the node label as constant $(x_2, 1)$, $\tilde{l}(x_2, 1) = 4\tilde{0}$, $\tilde{p} = (x_2, 1)$.

Step 5. $p \neq t'$, turn to the *step 2*.

Seventh iteration

Step 2. Change the labels for all nodes with temporary labels
$\Gamma(p) = \Gamma(x_2, 1) = \{(x_4, 2)\}$—the node $(x_4, 2)$ has temporary label. We obtain from (3.22).

$$\tilde{l}(x_4, 2) = \min[\infty, 4\tilde{0} + 4\tilde{0}] = 8\tilde{0}.$$

Step 3. Define the minimum value for nodes

$$(x_2, 0), (x_2, 3), (x_3, 0), (x_3, 1), (x_3, 2), (x_3, 3), (x_4, 0), (x_4, 1), (x_4, 2), (x_4, 3),$$
$$(x_5, 0), (x_5, 1), (x_5, 2), (x_5, 3), t',$$

as $\min[\infty, \infty, \infty, \infty, \infty, 4\tilde{0}, \infty, \infty, 8\tilde{0}, \infty, \infty, \infty, \infty, 10\tilde{5}, \infty] = 4\tilde{0}$.   The   label $(x_3, 3)$ has the minimum value.

Step 4. Consider the node label $(x_3, 3)$ as constant, $\tilde{l}(x_3, 3) = 4\tilde{0}, \tilde{p} = (x_3, 3)$.
Step 5. $p \neq t'$, turn to the *step 2*.

Eighth iteration

Step 2. Change the labels for all nodes with temporary labels:

$$\Gamma(p) = \Gamma(x_3, 3) = \{\varnothing\}.$$

Step 3. Define the minimum value for nodes

$$(x_2, 0), (x_2, 3), (x_3, 0), (x_3, 1), (x_3, 2), (x_4, 0), (x_4, 1), (x_4, 2), (x_4, 3),$$
$$(x_5, 0), (x_5, 1), (x_5, 2), (x_5, 3), t',$$

as $\min[\infty, \infty, \infty, \infty, \infty, \infty, \infty, 8\tilde{0}, \infty, \infty, \infty, \infty, 10\tilde{5}, \infty] = 8\tilde{0}$. The label $(x_4, 2)$ has the minimum value.

Step 4. Consider the node label $(x_4, 2)$ as constant, $\tilde{l}(x_4, 2) = 8\tilde{0}, \tilde{p} = (x_4, 2)$.
Step 5. $p \neq t'$, turn to the *step 2*.

Ninth iteration

Step 2. Change the labels for all nodes with temporary labels:
$\Gamma(p) = \Gamma(x_4, 2) = \{(x_5, 3)\}$—the node $(x_5, 3)$ has constant label. Obtain from (3.22):
$\tilde{l}(x_5, 3) = \min[10\tilde{5}, 8\tilde{0} + 3\tilde{0}] = 10\tilde{5}$.
Step 3. Define the minimum value for nodes

$$(x_2, 0), (x_2, 3), (x_3, 0), (x_3, 1), (x_3, 2), (x_4, 0), (x_4, 1), (x_4, 3),$$
$$(x_5, 0), (x_5, 1), (x_5, 2), (x_5, 3), t'$$

as $\min[\infty, \infty, \infty, \infty, \infty, \infty, \infty, \infty, \infty, \infty, \infty, 10\tilde{5}, \infty] = 10\tilde{5}$. The label $(x_5, 3)$ has the minimum value.

Step 4. Consider the node label $(x_5, 3)$ as constant, $\tilde{l}(x_5, 3) = 10\tilde{5}, \tilde{p} = (x_5, 3)$.
Step 5. $p \neq t'$, turn to the *step 2*.

Tenth iteration

Step 2. Change the labels for all nodes with temporary labels:
$\Gamma(p) = \Gamma(x_5, 3) = \{t'\}$—the node t' has temporary label. Obtain from (3.22):

$$\tilde{l}(t') = \min[\infty, 10\tilde{5} + tilde0] = 10\tilde{5}.$$

Step 3. Define the minimum value for nodes

$$(x_2, 0), (x_2, 3), (x_3, 0), (x_3, 1), (x_3, 2), (x_4, 0), (x_4, 1), (x_4, 3), (x_5, 0), (x_5, 1), (x_5, 2), t'$$

as $\min[\infty, \infty, \infty, \infty, \infty, \infty, \infty, \infty, \infty, \infty, \infty, 10\tilde{5}] = 10\tilde{5}$. The label has the minimum value t'.

Step 4. Consider the label of the node t' as constant, $\tilde{l}(t') = 10\tilde{5}, p \neq t'$.

Step 5. $p \neq t'$, therefore, we obtain the path of the minimum cost $\tilde{l}(t') = 10\tilde{5}$. Find this path according to the equation:

$$\tilde{l}(x_i', \varsigma) + \tilde{c}(x_i', x_i, \varsigma, \theta) = \tilde{l}(x_i, \theta). \tag{3.23}$$

In (3.23) (x_i', ς)—the node, that precedes the node (x_i, θ) in the minimum cost path from s' to (x_i, θ).

Determine the node, precedes the node t' in the minimum cost path from s' to t':

$$\tilde{l}(x_5, 0) + \tilde{c}(x_5, t', 0, \forall) \neq \tilde{l}(t'), \quad \text{as} \quad \infty + \tilde{0} \neq 10\tilde{5},$$
$$\tilde{l}(x_5, 1) + \tilde{c}(x_5, t', 1, \forall) \neq \tilde{l}(t'), \quad \text{as} \quad \infty + \tilde{0} \neq 10\tilde{5},$$
$$\tilde{l}(x_5, 2) + \tilde{c}(x_5, t', 2, \forall) \neq \tilde{l}(t'), \quad \text{as} \quad \infty + \tilde{0} \neq 10\tilde{5},$$
$$\tilde{l}(x_5, 3) + \tilde{c}(x_5, t', 3, \forall) \neq \tilde{l}(t'), \quad \text{as} \quad 10\tilde{5} + \tilde{0} \neq 10\tilde{5}.$$

Obtain the path $(x_5, 3)t'$.

Determine the node, precedes the node $(x_5, 3)$ in the minimum cost path from s' to t':

$$\tilde{l}(x_2, 2) + \tilde{c}(x_2, x_5, 2, 3) = \tilde{l}(x_5, 3) \quad \text{as} \quad 3\tilde{0} + 7\tilde{5} = 10\tilde{5}.$$
$$\tilde{l}(x_4, 2) + \tilde{c}(x_4, x_5, 2, 3) \neq \tilde{l}(x_5, 3) \quad \text{as} \quad 8\tilde{0} + 3\tilde{0} \neq 10\tilde{5},$$

Therefore, the path is obtained $(x_2, 2), (x_5, 3), t'$.

Determine the node, precedes the node $(x_2, 2)$ in the minimum cost path from s' to t':

$$\tilde{l}(x_1, 1) + \tilde{c}(x_1, x_2, 1, 2) = \tilde{l}(x_2, 2) \quad \text{as} \quad \tilde{0} + 3\tilde{0} = 3\tilde{0}.$$

Therefore, the path is obtained $(x_1, 1), (x_2, 2), (x_5, 3), t'$.

Determine the node, that precedes the node $(x_1, 1)$ in the minimum cost path from s' to t':

$$\tilde{l}(s') + \tilde{c}(s', x_1, \forall, 1) = \tilde{l}(x_1, 1) \quad \text{as} \quad \tilde{0} + \tilde{0} = \tilde{0}.$$

Therefore, the path is obtained $s', (x_1, 1), (x_2, 2), (x_5, 3), t'$.

Step 8. Determine new values of node potentials:

$$\pi(s') = \tilde{0} - \tilde{0} = \tilde{0},$$
$$\pi(x_1, 0) = \tilde{0} - \tilde{0} = \tilde{0},$$
$$\pi(x_1, 1) = \tilde{0} - \tilde{0} = \tilde{0},$$
$$\pi(x_1, 2) = \tilde{0} - \tilde{0} = \tilde{0},$$
$$\pi(x_1, 3) = \tilde{0} - \tilde{0} = \tilde{0},$$
$$\pi(x_2, 0) = \tilde{0} - 10\tilde{5} = -10\tilde{5},$$
$$\pi(x_2, 1) = \tilde{0} - 4\tilde{0} = -4\tilde{0},$$
$$\pi(x_2, 2) = \tilde{0} - 3\tilde{0} = -3\tilde{0},$$
$$\pi(x_2, 3) = \tilde{0} - 10\tilde{5} = -10\tilde{5},$$
$$\pi(x_3, 0) = \tilde{0} - 10\tilde{5} = -10\tilde{5},$$
$$\pi(x_3, 1) = \tilde{0} - 10\tilde{5} = -10\tilde{5},$$
$$\pi(x_3, 2) = \tilde{0} - 10\tilde{5} = -10\tilde{5},$$
$$\pi(x_3, 3) = \tilde{0} - 4\tilde{0} = -4\tilde{0},$$
$$\pi(x_4, 0) = \tilde{0} - 10\tilde{5} = -10\tilde{5},$$
$$\pi(x_4, 1) = \tilde{0} - 10\tilde{5} = -10\tilde{5},$$
$$\pi(x_3, 2) = \tilde{0} - 8\tilde{0} = -8\tilde{0},$$
$$\pi(x_4, 3) = \tilde{0} - 10\tilde{5} = -10\tilde{5},$$
$$\pi(x_5, 0) = \tilde{0} - 10\tilde{5} = -10\tilde{5},$$
$$\pi(x_5, 1) = \tilde{0} - 10\tilde{5} = -10\tilde{5},$$
$$\pi(x_5, 2) = \tilde{0} - 10\tilde{5} = -10\tilde{5},$$
$$\pi(x_5, 3) = \tilde{0} - 10\tilde{5} = -10\tilde{5},$$
$$\pi(t') = \tilde{0} - 10\tilde{5} = -10\tilde{5}.$$

Step 9. Push $\tilde{\delta}_p^{\mu} = \min\{\tilde{u}(\widetilde{P}_p^{\mu}), \tilde{e}(s'), -\tilde{e}(t')\}$, where $\tilde{u}(\widetilde{P}_p^{\mu}) = \min[\infty, 2\tilde{0}, 2\tilde{5}, \infty] = 20$, i.e. $\tilde{\delta}_p^{\mu} = \min\{20, 20, 20\} - 20$ flow units.

Step 10. Update flow values in $\widetilde{G}_p$.

The flow $\tilde{\xi}(x_i, x_j, \theta, \vartheta) = \tilde{0}$ turns to $2\tilde{0}$. Build a graph with the new flow value, as shown in Fig. 3.28, and turn to the step 4.

Step 4. Check, if there is an exceed in the node s'. As $e(s') = 2\tilde{0} + \tilde{0} - 2\tilde{0} = \tilde{0}$, the node s' is balanced. Therefore, the given flow of the minimum cost is found and turn to the step 11.

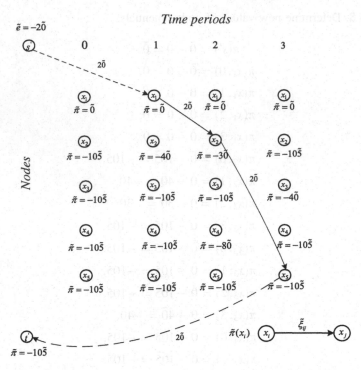

Fig. 3.28 Graph $\widetilde{G}_p$ with the flow of 20 units and node potentials

Step 11. Define the cost of the flow, which is equal to $2\tilde{0}$ units, as: $2\tilde{0} \cdot 3\tilde{0} + 2\tilde{0} \cdot 7\tilde{5} = 210\tilde{0}$ conventional units in $\widetilde{G}_p$. Turning to dynamic graph $\widetilde{G}$ from the «time-expanded» static graph $\widetilde{G}_p$, we can conclude, that given dynamic flow for 3 time periods, equals $2\tilde{0}$ units is equal to the flow leaving the «node-time» pairs $(x_1, 1)$ and entering the «node-time» pairs $(x_5, 3)$, which are defined by the path $x_1 \rightarrow x_2 \rightarrow x_5$, with the flow sending at the time period $\theta = 1$ and entering the sink at the moment $\theta = 3$.

Define uncertainty borders of the obtained fuzzy number 20, corresponded to the given flow in the graph $\widetilde{G}$. The found result is between two basic neighboring values of fuzzy arc capacities: 15 with the left deviation $l_1^L = 5$, right deviation $-l_1^R = 5$ и 31 with the left deviation $l_2^L = 8$, right deviation—$l_2^R = 7$, presented in the form of the fuzzy triangular numbers. According to the (2.4):

$$l^L = \frac{(a_2 - a')}{(a_2 - a_1)} \times l_1^L + \left(1 - \frac{(a_2 - a')}{(a_2 - a_1)}\right) \times l_2^L$$

$$= \frac{(31 - 20)}{(31 - 15)} \times 5 + \left(1 - \frac{(31 - 20)}{(31 - 15)}\right) \times 8$$

$$= 5.94 \approx 6,$$

Fig. 3.29 Given flow value in the form of the fuzzy trapezoidal number of (20,6,6) units

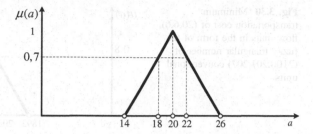

$$l^R = \frac{(a_2 - a')}{(a_2 - a_1)} \times l_1^R + \left(1 - \frac{(a_2 - a')}{(a_2 - a_1)}\right) \times l_2^R$$

$$= \frac{(31 - 20)}{(31 - 15)} \times 5 + \left(1 - \frac{(31 - 20)}{(31 - 15)}\right) \times 7$$

$$= 5.63 \approx 6.$$

Therefore, we can present the value of the given flow in the form of the fuzzy triangular number (20,6,6), as shown in Fig. 3.29.

According to Fig. 3.29 the flow (20,6,6) of the minimum cost with the degree of confidence 1 will equal to 20 units, with the degree of confidence 0,7 will be within the interval [18,22], but anyway, the minimum cost flow will be no less than 14 units and no more than 26 units.

Represent the minimum transportation cost of the (20,6,6) flow units, equal $210\tilde{0}$ conventional units in the form of the fuzzy triangular number.

Let us define deviation borders of the obtained fuzzy number of $210\tilde{0}$ conventional units. Found result is between two basic neighboring values of fuzzy transmission costs: $184\tilde{0}$ with the left deviation $l_1^L = 190$, right deviation—$l_1^R = 190$ and $243\tilde{0}$ with the left deviation $l_2^L = 230$, right deviation—$l_2^R = 230$. According to (2.4) we obtain:

$$l^L = \frac{(a_2 - a')}{(a_2 - a_1)} \times l_1^L + \left(1 - \frac{(a_2 - a')}{(a_2 - a_1)}\right) \times l_2^L$$

$$= \frac{(2430 - 2100)}{(2430 - 1840)} \times 190 + \left(1 - \frac{(2430 - 2100)}{(2430 - 1840)}\right) \times 230$$

$$= 207.4 \approx 207,$$

$$l^R = \frac{(a_2 - a')}{(a_2 - a_1)} \times l_1^R + \left(1 - \frac{(a_2 - a')}{(a_2 - a_1)}\right) \times l_2^R$$

$$= \frac{(2430 - 2100)}{(2430 - 1840)} \times 190 + \left(1 - \frac{(2430 - 2100)}{(2430 - 1840)}\right) \times 230$$

$$= 207.4 \approx 207.$$

Fig. 3.30 Minimum transportation cost of (20,6,6) flow units in the form of the fuzzy triangular number of (2100,207,207) conventional units

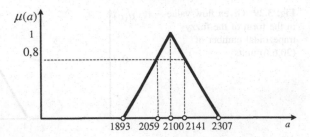

Therefore, we can present the minimum transportation flow cost in the form of the triangular fuzzy number (2100,207,207), as shown in Fig. 3.30.

According to Fig. 3.30 minimum transportation cost of (20,6,6) flow units with the degree of confidence 0,8 will be within the interval of [2059,2141] conventional units, but anyway, the minimum transmission cost of (20,6,6) flow units will be no less than 1893 and no more than 2307 conventional units.

Minimum cost maximum flow finding in dynamic network with transit fuzzy arc capacities and transmission costs.

Important task arising in dynamic networks is the task of the maximum dynamic flow finding of the minimum cost.

The problem statement of the maximum dynamic flow finding of the minimum cost in crisp conditions and in terms of stationary-dynamic graphs and literature review on this subject are presented in the first chapter of this book. This task on the fuzzy dynamic graphs, i.e. those, which parameters can vary over time wasn't considered in fuzzy conditions.

Other words, it is necessary to determine minimum transmission cost of the maximum flow in fuzzy graph $\widetilde{G} = (X, \widetilde{A})$ with transit arcs along the arcs such a way, that the last flow unit would enter the sink at the moment p. In this case the problem statement of this task is:

$$Minimize \sum_{\theta=0}^{p} \sum_{(x_i,x_j)\in \widetilde{A}} \tilde{c}_{ij}(\theta)\tilde{\xi}_{ij}(\theta), \quad \forall(x_i,x_j)\in \widetilde{A},\ \theta\in T, \qquad (3.24)$$

$$\sum_{\theta=0}^{p} \left(\sum_{x_j\in\Gamma(x_i)} \tilde{\xi}_{ij}(\theta) - \sum_{x_j\in\Gamma^{-1}(x_i)} \tilde{\xi}_{ij}(\theta - \tau_{ji}(\theta)) \right) = \tilde{v}(p), \quad x_i = s, \qquad (3.25)$$

$$\sum_{\theta=0}^{p} \left(\sum_{x_j\in\Gamma(x_i)} \tilde{\xi}_{ij}(\theta) - \sum_{x_j\in\Gamma^{-1}(x_i)} \tilde{\xi}_{ij}(\theta - \tau_{ji}(\theta)) \right) = \tilde{0}, \quad x_i \neq s,t;\ \theta\in T, \qquad (3.26)$$

$$\sum_{\theta=0}^{p} \left(\sum_{x_j \in \Gamma(x_i)} \tilde{\xi}_{ij}(\theta) - \sum_{x_j \in \Gamma^{-1}(x_i)} \tilde{\xi}_{ij}(\theta - \tau_{ji}(\theta)) \right) = -\tilde{v}(p), \quad x_i = t, \quad (3.27)$$

$$\tilde{0} \le \tilde{\xi}_{ij}(\theta) \le \tilde{u}_{ij}(\theta), \quad \forall (x_i, x_j) \in \tilde{A}, \ \theta \in T. \quad (3.28)$$

It is necessary to develop algorithm of the maximum flow of the minimum cost in dynamic network in terms of partial uncertainty. Introduce the *rule 3.9* of constructing the fuzzy residual network in the "time-expanded" graph.

Rule 3.9 of constructing fuzzy residual network in the "time-expanded" graph for the minimum cost maximum flow finding in dynamic network with fuzzy transit arc capacities and transmission costs.

Fuzzy residual network $\widetilde{G}_p^\mu = (X_p^\mu, \widetilde{A}_p^\mu)$ is constructed according to the "time-expanded" network $\widetilde{G}_p = (X_p, \widetilde{A}_p)$ depending on the flow values $\tilde{\xi}(x_i, x_j, \theta, \vartheta = \theta + \tau_{ij}(\theta))$, going along the network as follows: each arc in the residual network $\widetilde{G}_p^\mu$, connecting the "node-time" pair (x_i^μ, θ) with the "node-time" pair (x_j^μ, ϑ) with the flow $\tilde{\xi}(x_i, x_j, \theta, \vartheta)$ going at the time $\theta \in T$ has residual network $\tilde{u}^\mu(x_i, x_j, \theta, \vartheta) = \tilde{u}(x_i, x_j, \theta, \vartheta) - \tilde{\xi}(x_i, x_j, \theta, \vartheta)$, transmission cost of the one flow unit $\tilde{c}^\mu(x_i, x_j, \theta, \vartheta) = \tilde{c}(x_i, x_j, \theta, \vartheta)$ with transit time $\tau^\mu(x_i, x_j, \theta, \vartheta) = \tau(x_i, x_j, \theta, \vartheta)$ and reverse arc, connecting (x_j^μ, ϑ) with (x_i^μ, θ) with residual arc capacity $\tilde{u}^\mu(x_j, x_i, \vartheta, \theta) = \tilde{\xi}(x_i, x_j, \theta, \vartheta)$, cost $\tilde{c}^\mu(x_j, x_i, \vartheta, \theta) = \tilde{c}(x_i, x_j, \theta, \vartheta)$ and transit time $\tau^\mu(x_j, x_i, \vartheta, \theta) = -\tau(x_i, x_j, \theta, \vartheta)$.

Algorithm of minimum cost maximum flow finding in dynamic network with fuzzy transit arc capacities and transmission costs.

Consider formal algorithm of the maximum flow finding in dynamic network with fuzzy transit arc capacities and transmission costs in terms of partial uncertainty, presented by the model (3.24)–(3.28).

Step 1. Turn from the given fuzzy dynamic graph $\widetilde{G}$ to the "time-expanded" fuzzy static graph $\widetilde{G}_p$ according to time-expanding of initial dynamic graph for given amount of time periods by making particular copy of each node $x_i \in X$ at each considered time period $\theta \in T$ according to the *rule 3.7*. It is necessary to calculate maximum flow of the minimum cost, leaving the group of sources at all time periods and entering the group of sinks at all time periods not later than p. Introduce artificial source s' and sink t' and connect s' by arcs with true source and t' with each true sink. Artificial arcs going from artificial nodes have infinite arc capacity.

Step 2. Build fuzzy residual network $\widetilde{G}_p^\mu$ depending on flow values going along the arcs of the graph. Fuzzy residual network $\widetilde{G}_p = (X_p, \widetilde{A}_p)$ is constructed by the "time-expanded" network $\widetilde{G}_p$ depending on the flow values, starting from zero flows according the *rule 3.9*. Initially residual network coincides with the "time-expanded" static graph.

Step 3. Search the minimum cost path $\widetilde{P}_p^\mu$ from the artificial source s' to the artificial sink t' in constructed "time-expanded" graph of the fuzzy residual network $\widetilde{G}_p^\mu$.

3.1. If the path $\widetilde{P}_p^\mu$ is found, turn to the step 4.

3.2. If the path is failed to find, the maximum flow $\widetilde{\xi}(x_i, x_j, \theta, \vartheta) + \widetilde{\delta}_p^\mu \times \widetilde{P}_p^\mu = \widetilde{v}(p)$ of the minimum cost $\widetilde{c}\left(\widetilde{\xi}(x_i, x_j, \theta, \vartheta) + \widetilde{\delta}_p^\mu \times \widetilde{P}_p^\mu\right)$ is obtained in the "time-expanded" static fuzzy graph from s' to t' and turn to the step 6.

Step 4. Push maximum amount of flow units depending on the arc in the residual network with minimum residual arc capacity $\widetilde{\delta}_p^\mu = \min[\widetilde{u}^\mu(x_i, x_j, \theta, \vartheta)], [(x_i, \theta), (x_j, \vartheta)] \in \widetilde{P}_p^\mu$.

Step 5. Update flow values in $\widetilde{G}_p$: replace the flow $\widetilde{\xi}(x_j, x_i, \theta, \vartheta)$ by $\widetilde{\xi}(x_j, x_i, \theta, \vartheta) - \widetilde{\delta}_p^\mu$ along the corresponding arcs, going from (x_j, θ) to (x_i, ϑ) from $\widetilde{G}_p$ for arcs, connecting node-time pair (x_i^μ, ϑ) with (x_j^μ, θ) in $\widetilde{G}_p^\mu$, such as $((x_i^\mu, \vartheta), (x_j^\mu, \theta)) \notin \widetilde{A}_p, ((x_i^\mu, \vartheta), (x_j^\mu, \theta)) \in \widetilde{A}_p^\mu$, and replace the flow $\widetilde{\xi}(x_i, x_j, \theta, \vartheta)$ by $\widetilde{\xi}(x_i, x_j, \theta, \vartheta) + \widetilde{\delta}_p^\mu$ along the corresponding arcs, going from (x_i, θ) to (x_j, ϑ) from $\widetilde{G}_p$ for arcs, connecting node-time pair (x_i^μ, θ) with (x_j^μ, ϑ) in $\widetilde{G}_p^\mu$, such as $((x_i^\mu, \theta), (x_j^\mu, \vartheta)) \in \widetilde{A}_p, ((x_i^\mu, \theta), (x_j^\mu, \vartheta)) \in \widetilde{A}_p^\mu$. Replace the flow value in $\widetilde{G}_p$: $\widetilde{\xi}(x_i, x_j, \theta, \vartheta)$ by $\widetilde{\xi}(x_i, x_j, \theta, \vartheta) + \widetilde{\delta}_p^\mu \widetilde{P}_p^\mu$ and turn to the step 2.

Step 6. If the maximum flow $\widetilde{\xi}(x_i, x_j, \theta, \vartheta) + \widetilde{\delta}_p^\mu \times \widetilde{P}_p^\mu = \widetilde{v}(p)$ of the minimum cost $\widetilde{c}(\widetilde{\xi}(x_i, x_j, \theta, \vartheta) + \widetilde{\delta}_p^\mu \times \widetilde{P}_p^\mu)$ from the artificial source to the artificial sink, defined by the set of paths $\widetilde{P}_p^\mu$ is found, turn to the initial dynamic graph $\widetilde{G}$ as follows: reject the artificial nodes s', t' and arcs, connecting them with other nods.

Thus, the maximum flow of the value $\widetilde{v}(p)$, which is equivalent to the flow from the sources (the initial node expanded to the p intervals) to the sinks (terminal node expanded to the p intervals) is obtained in $\widetilde{G}_p$ after deleting artificial nodes and each path, connecting nodes (s, ς) and $(t, \psi = \varsigma + \tau_{st}(\varsigma)), \psi \in T$, with the flow $\widetilde{\xi}_{st}(s, t, \theta, \varsigma)$ coming along, corresponds to the flow $\widetilde{\xi}_{st}(\theta)$ of the cost $\widetilde{c}(\widetilde{\xi}_{st}(\theta))$.

Consider numerical example, realized the minimum cost maximum flow finding algorithm in fuzzy dynamic graph.

Numerical example 8
Let the network, which is the part of the railway map, is presented in the form of the fuzzy directed graph, obtained from GIS «ObjectLand» (Fig. 3.11). Fuzzy arc capacities, transmission costs and transit times, depending on the flow departure time are presented in the form of Tables 3.6, 3.7 and 3.8. It is necessary to find the minimum cost of the maximum amount of flow for 3 time periods and present result as fuzzy triangular number, if there are basic values of arc capacities and costs, set as fuzzy triangular number, as shown in Figs. 2.27–2.30.

Step 1. Turn to the "time-expanded" static graph from the initial dynamic graph according to the *rule 3.7*, as shown in Fig. 3.31. Introduce artificial source and sink and connect them by arcs with other nodes, as shown in Fig. 3.32.

Step 2. Fuzzy residual network $\widetilde{G}_p^\mu$ coincides with network, presented in Fig. 3.32 at this step.

Step 3. Search the minimum cost path $\widetilde{P}_1^\mu$ from s' to t' in $\widetilde{G}_p^\mu$: $\widetilde{P}_1^\mu = s', (x_1, 1), (x_2, 2), (x_5, 3), t'$.

Step 4. Push $\widetilde{P}_1^\mu = s', (x_1, 1), (x_2, 2), (x_5, 3), t'$ $\widetilde{\delta}_p^\mu = \min[\infty, 2\widetilde{0}, 2\widetilde{5}, \infty] = 2\widetilde{0}$ flow units along this path.

Step 5. Update flow values in $\widetilde{G}_p$. The flow $\widetilde{\xi}(x_i, x_j, \theta, \vartheta) = \widetilde{0}$ turns to $\widetilde{\xi}(x_i, x_j, \theta, \vartheta) = \widetilde{0} + 2\widetilde{0} = 2\widetilde{0}$ units. Build a graph with the new flow value, as shown in Fig. 3.33 turn to the step 2.

Step 2. Build fuzzy residual network $\widetilde{G}_p^\mu$ according to the flow value, passing along the arcs of the graph in Fig. 3.33, as shown in Fig. 3.34.

Step 3. Search a minimum cost path $\widetilde{P}_2^\mu$ from s' to t' in $\widetilde{G}_p^\mu$: $\widetilde{P}_2^\mu = s', (x_1, 0), (x_2, 1), (x_4, 2), (x_5, 3), t'$.

Step 4. Pass $\widetilde{\delta}_p^\mu = \min[\infty, 1\widetilde{0}, 2\widetilde{0}, 3\widetilde{0}, \infty] = 1\widetilde{0}$ flow units along the minimum cost path $\widetilde{P}_2^\mu = s', (x_1, 0), (x_2, 1), (x_4, 2), (x_5, 3), t'$.

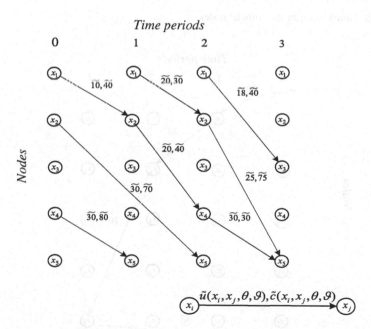

Fig. 3.31 «Time-expanded» graph $\widetilde{G}_p$

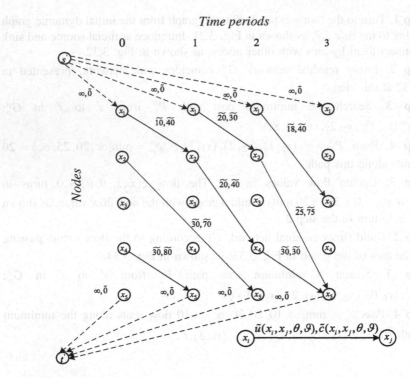

Fig. 3.32 Graph $\widetilde{G}_p$ with the artificial nodes

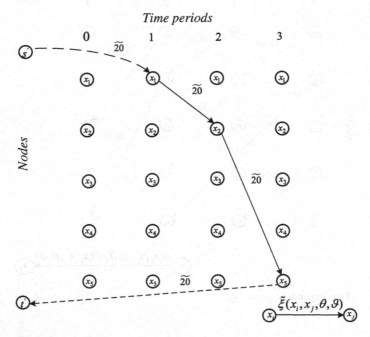

Fig. 3.33 Graph $\widetilde{G}_p$ with the flow of $\widetilde{20}$ units

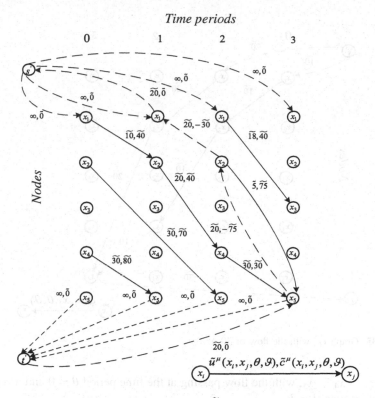

Fig. 3.34 «Time-expanded» fuzzy residual network $\widetilde{G}_p$ for the graph in Fig. 3.33

Step 5. Update flow values in $\widetilde{G}_p$. The flow $\widetilde{\xi}(x_i, x_j, 0, \vartheta) = 2\widetilde{0}$ turns to $\widetilde{\xi}(x_i, x_j, 0, \vartheta) = 2\widetilde{0} + 1\widetilde{0} = 3\widetilde{0}$ units. Build a graph with the new flow value, as shown in Fig. 3.35 and turn to the step 2.

Step 2. Build fuzzy residual network $\widetilde{G}_p^\mu$ according to the flow values, going along the arcs of the graph in Fig. 3.35, as shown in Fig. 3.36.

Step 3. Search the minimum cost path $\widetilde{P}_3^\mu$ from s' to t' in $\widetilde{G}_p^\mu$. This path doesn't exist, therefore, we find the maximum flow of the minimum cost in $\widetilde{G}_p$, which represented in Fig. 3.37.

Maximum flow is 30 units of the minimum cost $1\widetilde{0} \cdot 4\widetilde{0} + 1\widetilde{0} \cdot 4\widetilde{0} + 1\widetilde{0} \cdot 3\widetilde{0} + 2\widetilde{0} \cdot 3\widetilde{0} + 2\widetilde{0} \cdot 7\widetilde{5} = 320\widetilde{0}$ conventional units.

Step 6. Turning to the dynamic graph $\widetilde{G}$ from the "time-expanded" static graph $\widetilde{G}_p$, we can conclude, that maximum dynamic flow for 3 time periods equals 30 flow units is equal to the flow, leaving the "node-time" pairs $(x_1, 0)$ and $(x_1, 1)$ and entering the "node-time" pair $(x_5, 3)$, which is defined by paths $x_1 \rightarrow x_2 \rightarrow x_5$, with the flow passing at the time $\theta = 1$ and arriving at the sink at time period $\theta = 3$, и

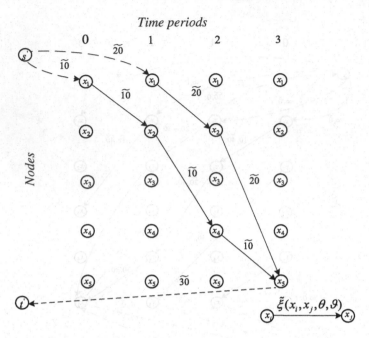

Fig. 3.35 Graph $\widetilde{G}_p$ with the flow of $3\tilde{0}$ units

$x_1 \rightarrow x_2 \rightarrow x_4 \rightarrow x_5$, with the flow passing at the time period $\theta = 0$ and arriving at the sink at time $\theta = 3$.

Let us define borders of uncertainty of the obtained fuzzy number 30, corresponded to the maximum flow in $\widetilde{G}$. The found result is between two basic fuzzy values of arc capacities: 15 with the left deviation $l_1^L = 5$, right deviation—$l_1^R = 5$ and 31 with the left deviation $l_2^L = 8$, right deviation—$l_2^R = 7$, presented as fuzzy triangular numbers. According to (2.4) we obtain:

$$l^L = \frac{(a_2 - a')}{(a_2 - a_1)} \times l_1^L + \left(1 - \frac{(a_2 - a')}{(a_2 - a_1)}\right) \times l_2^L$$

$$= \frac{(31 - 30)}{(31 - 15)} \times 5 + \left(1 - \frac{(31 - 30)}{(31 - 15)}\right) \times 8$$

$$= 7.81 \approx 8$$

$$l^R = \frac{(a_2 - a')}{(a_2 - a_1)} \times l_1^R + \left(1 - \frac{(a_2 - a')}{(a_2 - a_1)}\right) \times l_2^R$$

$$= \frac{(31 - 30)}{(31 - 15)} \times 5 + \left(1 - \frac{(31 - 30)}{(31 - 15)}\right) \times 7$$

$$= 6.87 \approx 7.$$

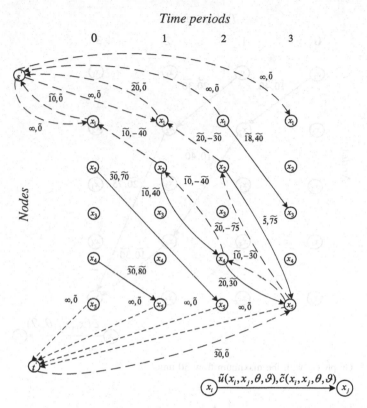

Fig. 3.36 «Time-expanded» fuzzy residual network $\widetilde{G}_p$ for the graph in Fig. 3.35

Therefore, we can represent the maximum flow value in the form of the fuzzy triangular number (30,8,7), as shown in Fig. 3.38.

According to Fig. 3.38 the flow (30,8,7) of the minimum cost with the degree of confidence 0,7 will be within the interval [28,32], but anyway the minimum cost flow will no less than 22 units and no more than 37 units.

Let us define minimum transportation cost of (30,8,7) units equals $320\widetilde{0}$ conventional units in the form of the fuzzy triangular number.

Let us define borders of uncertainty of the obtained fuzzy number $320\widetilde{0}$ conventional units. Found result is between two basic neighboring values of fuzzy transmission costs: 3100 with the left deviation $l_1^L = 300$, right deviation—$l_1^R = 350$ and $430\widetilde{0}$ with the left deviation $l_2^L = 450$, right deviation—$l_2^R = 470$. According to (2.4) we obtain:

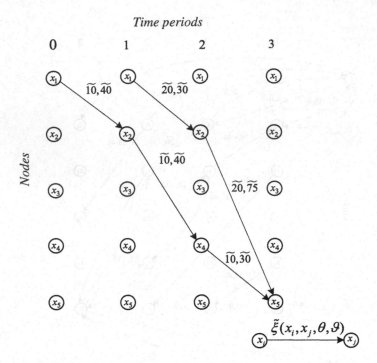

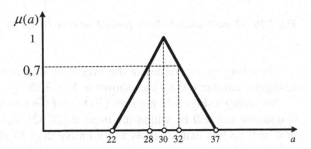

Fig. 3.37 Graph $\widetilde{G}_p$ with the maximum flow 30 units

Fig. 3.38 Maximum flow in
the form of the fuzzy
trapezoidal number of
(30,8,7) units

$$l^L = \frac{(a_2 - a')}{(a_2 - a_1)} \times l_1^L + \left(1 - \frac{(a_2 - a')}{(a_2 - a_1)}\right) \times l_2^L$$

$$= \frac{(4300 - 3200)}{(4300 - 3100)} \times 300 + \left(1 - \frac{(4300 - 3200)}{(4300 - 3100)}\right) \times 450$$

$$= 311$$

Fig. 3.39 Minimum
transmission cost of (30,8,7)
flow units in the form of the
fuzzy triangular number of
(3200,311,358) conventional
units

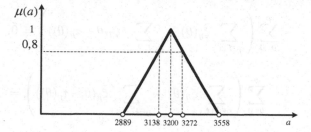

$$l^R = \frac{(a_2 - a')}{(a_2 - a_1)} \times l_1^R + \left(1 - \frac{(a_2 - a')}{(a_2 - a_1)}\right) \times l_2^R$$

$$= \frac{(4300 - 3200)}{(4300 - 3100)} \times 350 + \left(1 - \frac{(4300 - 3200)}{(4300 - 3100)}\right) \times 470$$

$$= 358.43 \approx 358.$$

Therefore, we can represent the minimum flow transportation cost in the form of
the fuzzy triangular number (3200,311,358), as shown in Fig. 3.39.

According to Fig. 3.39 the minimum transportation cost of (30,8,7) units with
the degree of confidence 0,8 will be within the interval of [3138,3272] conventional
units, but anyway, the minimum transportation cost of (30,7,6) units will be no less
than 2889 and no more than 3558 conventional units.

3.5 Minimum Cost Flow Finding in Dynamic Network with Fuzzy Nonzero Lower, Upper Flow Bounds and Transmission Costs

Let us consider situation, when it is necessary to transport given flow value from the
source to the sink in the network in terms of fuzziness and transit time along the arc
and transit parameters and given minimum values of transportation profitability, set
as lower flow bounds. Then we turn to the minimum cost flow problem statement
with lower flow bounds in dynamic network [15]. The model of the task is:

$$Minimize \sum_{\theta=0}^{p} \sum_{(x_i,x_j) \in \tilde{A}} \tilde{c}_{ij}(\theta)\tilde{\xi}_{ij}(\theta), \quad \forall(x_i,x_j) \in \tilde{A}, \ \theta \in T, \qquad (3.29)$$

$$\sum_{\theta=0}^{p} \left(\sum_{x_j \in \Gamma(x_i)} \tilde{\xi}_{ij}(\theta) - \sum_{x_j \in \Gamma^{-1}(x_i)} \tilde{\xi}_{ij}(\theta - \tau_{ji}(\theta))\right) = \tilde{p}(p), \quad x_i = s, \qquad (3.30)$$

$$\sum_{\theta=0}^{p} \left(\sum_{x_j \in \Gamma(x_i)} \tilde{\xi}_{ij}(\theta) - \sum_{x_j \in \Gamma^{-1}(x_i)} \tilde{\xi}_{ji}(\theta - \tau_{ji}(\theta)) \right) = \tilde{0}, \quad x_i \neq s, t; \ \theta \in T, \qquad (3.31)$$

$$\sum_{\theta=0}^{p} \left(\sum_{x_j \in \Gamma(x_i)} \tilde{\xi}_{ij}(\theta) - \sum_{x_j \in \Gamma^{-1}(x_i)} \tilde{\xi}_{ji}(\theta - \tau_{ji}(\theta)) \right) = -\tilde{\rho}(p), \quad x_i = t, \qquad (3.32)$$

$$\tilde{l}_{ij}(\theta) \leq \tilde{\xi}_{ij}(\theta) \leq \tilde{u}_{ij}(\theta), \quad \theta + \tau_{ij}(\theta) \leq p, \ \theta \in T. \qquad (3.33)$$

It is obvious, that the task, presented as the model (3.29)–(3.33) will have solution, when three conditions are satisfied:

1. The flows should satisfy the arc capacities and lower flow bounds constraints due to the condition (3.33).
2. Given flow value $\tilde{\rho}(p)$ should be no more than maximum feasible flow $\tilde{v}(p)$ in the initial graph, i.e. $\tilde{\rho}(p) \leq \tilde{v}(p)$.
3. Given flow value $\tilde{\rho}(p)$ should be no less than the sum of the lower flow bounds in the initial graph, i.e. $\tilde{\rho}(p) \geq \sum_{\tilde{l}(x_i, x_j, \vartheta, \theta) \neq \tilde{0}} \tilde{l}(x_i, x_j, \theta, \vartheta)$.

As this task was considered in the literature in crisp conditions and in terms of stationary-dynamic graphs, the necessity of developing method for the minimum cost dynamic flow finding with nonzero lower flow bounds and transit times along the arcs depend on the flow departure time appeared.

Thus, let is introduce the *rule 3.10* of turning to the time-expanded graph, corresponded to the initial graph, formulated in [15].

Rule 3.10 of constructing the "time-expanded" fuzzy graph corresponding to the initial dynamic graph for the minimum cost flow finding in dynamic network with fuzzy transit nonzero lower and upper flow, transmission costs.

Let $\widetilde{G}_p = (X_p, \widetilde{A}_p)$ represent fuzzy time-expanded static graph of the original dynamic fuzzy graph. The set of nodes X_p of the graph $\widetilde{G}_p$ is defined as $X_p = \{(x_i, \theta) : (x_i, \theta) \in X \times T\}$. The set of arcs $\widetilde{A}_p$ consists of arcs from each node-time pair $(x_j, \theta) \in X_p$ to every node-time pair $(x_j, \vartheta = \theta + \tau_{ij}, (\theta))$, where $x_j \in \Gamma(x_i)$ and $\theta + \tau_{ij}(\theta) \leq p$. Lower flow bounds $\tilde{l}(x_i x_j, \theta, \vartheta)$ joining (x_i, θ) with (x_j, ϑ) in $\widetilde{G}_p$ are equal to $\tilde{l}_{ij}(\theta)$. Fuzzy upper flow bounds $\tilde{u}(x_i, x_j, \theta, \vartheta)$ joining (x_i, θ) with (x_j, ϑ) are equal to $\tilde{u}_{ij}(\theta)$. Transmission costs of the one flow unit $\tilde{c}(x_i, x_j, \theta, \vartheta)$ joining (x_i, θ) with (x_j, ϑ) are equal to $\tilde{c}_{ij}(\theta)$. Transit times $\tau(x_i, x_j, \theta, \vartheta)$ joining (x_i, θ) with (x_j, ϑ) are equal to $\tau_{ij}(\theta)$.

Turn to the fuzzy time-expanded graph without lower flow bounds according to the *rule 3.11*, presented in [15].

Rule 3.11 of turning to the corresponding fuzzy graph without lower flow bounds from the "time-expanded" fuzzy graph for the minimum cost flow finding in dynamic network with fuzzy transit nonzero transit lower and upper flow bounds and costs.

Turn to the fuzzy graph $\widetilde{G}_p^* = (X_p^*, \widetilde{A}_p^*)$ without lower flow bounds from the "time-expanded" fuzzy graph $\widetilde{G}_p = (X_p, \widetilde{A}_p)$. Introduce the artificial source s^* and sink t^* and arcs connecting the node-time pair $(t, \forall \theta \in T)$ and $(s, \forall \theta \in T)$ with upper fuzzy flow bound $\tilde{u}^*(t, s, \forall \theta \in T, \forall \theta \in T) = \infty$, lower fuzzy flow bound $\tilde{l}^*(t, s, \forall \theta \in T, \forall \theta \in T) = \tilde{0}$, transmission cost $\tilde{c}^*(t, s, \forall \theta \in T, \forall \theta \in T) = \tilde{0}$ in the graph $\widetilde{G}_p$. It means that every node t at each time period from p is connected with every node s at all time periods in the graph $\widetilde{G}_p^*$ Introduce the following modification for each arc connecting the node-time pair (x_i, θ) with the node-time pair $(x_i, \vartheta = \theta + \tau_{ij}(\theta))$ with nonzero lower fuzzy flow bound $\tilde{l}(x_i, x_j, \theta, \vartheta) \neq \tilde{0}$: (1) reduce $\tilde{u}(x_i, x_j, \theta, \vartheta)$ to, $\tilde{u}^*(x_i, x_j, \theta, \vartheta) = \tilde{u}(x_i, x_j, \theta, \vartheta) - \tilde{l}(x_i, x_j, \theta, \vartheta), \tilde{l}(x_i, x_j, \theta, \vartheta)$ to $\tilde{0}$. (2) Introduce the arcs connecting s^* with (x_j, ϑ), and the arcs connecting t^* with (x_i, θ) with upper fuzzy flow bounds equal to lower fuzzy flow bounds $\tilde{u}^*(s^*, x_j, \forall, \vartheta) = \tilde{u}^*(x_i, t, \theta, \forall) = \tilde{l}(x_i, x_j, \theta, \vartheta)$ zero lower fuzzy flow bounds $\tilde{l}^*(s^*, x_j, \forall, \vartheta) = \tilde{l}^*(x_i, t, \theta, \forall) = \tilde{0}$, transmission costs $\tilde{c}^*(s^*, x_j, \forall, \vartheta) = \tilde{c}^*(x_i, t, \theta, \forall) = \tilde{0}$.

Introduce synthesized Definition 3.5 of the fuzzy residual network of the time-expanded graph for the minimum cost dynamic flow finding with nonzero lower flow bounds, presented in [15]. The criteria of the feasible flow finding in the time-expanded fuzzy graph, corresponding to the initial dynamic one is the maximum flow obtaining equals the sum of the lower flow bounds in the modified graph without lower flow bounds.

Definition 3.5 Fuzzy residual network $\widetilde{G}_p^{*\mu} = (X_p^{*\mu}, \widetilde{A}_p^{*\mu})$ of the time-expanded graph $\widetilde{G}_p^*$ for the maximum flow finding in dynamic network with nonzero lower and upper flow bounds—the network $\widetilde{G}_p^* = (X_p^*, \widetilde{A}_p^*)$ without lower flow bounds constructed depending the following rules: if the flow, going from the "node-time" pair (x_i^*, θ) to the node-time pair (x_i^*, ϑ) is less than arc capacity in $\widetilde{G}_p^*$, i.e. $\tilde{\xi}^*(x_i, x_j, \theta, \vartheta) < \tilde{u}^*(x_i, x_j, \theta, \vartheta)$, then include the corresponding arc, going from the node-time pair $(x_i^{*\mu}, \theta)$ to the node-time pair $(x_j^{*\mu}, \vartheta)$ with arc capacity $\tilde{u}^{*\mu}(x_i, x_j, \theta, \vartheta) = \tilde{u}^*(x_i, x_j, \theta, \vartheta) - \tilde{\xi}^{*\mu}(x_i, x_j, \theta, \vartheta)$ with transit time $\tau^{*\mu}(x_i, x_j, \theta, \vartheta) = \tau^*(x_i, x_j, \theta, \vartheta)$, transmission cost $\tilde{c}^{*\mu}(x_i, x_j, \theta, \vartheta) = \tilde{c}^*(x_i, x_j, \theta, \vartheta)$. If the flow going from the "node-time" pair (x_i^*, θ) to the node-time pair (x_i^*, ϑ) is more than $\tilde{0}$ in $\widetilde{G}_p^*$, i.e. $\tilde{\xi}^*(x_i, x_j, \theta, \vartheta) > \tilde{0}$, then include the corresponding arc, going from the node-time pair $(x_j^{*\mu}, \vartheta)$ to the node-time pair $(x_i^{*\mu}, \theta)$ in $\widetilde{G}_p^{*\mu}$ with arc capacity $\tilde{u}^\mu(x_j, x_i, \theta, \vartheta) = \tilde{\xi}^*(x_i, x_j, \theta, \vartheta)$ with transit time $\tau^{*\mu}(x_j, x_i, \vartheta, \theta) = -\tau^*(x_i, x_j, \theta, \vartheta)$ and transmission cost $\tilde{c}^{*\mu}(x_j, x_i, \vartheta, \theta) = -\tilde{c}^*(x_i, x_j, \theta, \vartheta)$.

Reverse transition to the time-expanded graph with the feasible flow, corresponding to the initial dynamic one from the time-expanded fuzzy graph without lower flow bounds with the maximum flow of the minimum cost is according to the *rule 3.5*. Then if the feasible flow in the time-expanded fuzzy graph is found, transform it, searching the minimum cost maximum flow according to the *rule 3.12*, presented in [15].

Rule 3.12 of the fuzzy residual network constructing with the feasible flow vector for the minimum cost flow finding in dynamic network with fuzzy nonzero transit lower and upper flow bounds and costs.

Rule 3.6 of the fuzzy residual network constructing with the feasible flow vector for the maximum flow finding in dynamic network with fuzzy nonzero transit lower and upper flow bounds.

If the flow, going from the node-time pair (x_i, θ) to the node-time pair (x_j, ϑ) is less, than arc capacity in $\widetilde{G}_p$, i.e. $\tilde{\xi}(x_i, x_j, \theta, \vartheta) < \tilde{u}(x_i, x_j, \theta, \vartheta)$, then include the corresponding arc (x_i^μ, θ) from the node-time pair to the node-time pair (x_j^μ, ϑ) in $\widetilde{G}_p^\mu(\tilde{\xi})$ with arc capacity $\tilde{u}^\mu(x_i, x_j, \theta, \vartheta) = \tilde{u}(x_i, x_j, \theta, \vartheta) - \tilde{\xi}(x_i, x_j, \theta, \vartheta)$, transit time $\tau^\mu(x_i, x_j, \theta, \vartheta) = \tau(x_i, x_j, \theta, \vartheta)$ and transmission cost $\tilde{c}^\mu(x_i, x_j, \theta, \vartheta) = \tilde{c}(x_i, x_j, \theta, \vartheta)$. If the flow, going from the node-time pair (x_i, θ) to the node-time pair (x_j, ϑ) is more than the lower flow bound in $\widetilde{G}_p$, i.e. $\tilde{\xi}(x_i, x_j, \theta, \vartheta) > \tilde{l}(x_i, x_j, \theta, \vartheta)$, then include the corresponding arc, going from the node-time pair (x_j^μ, ϑ) to the node-time pair (x_i^μ, θ) in $\widetilde{G}_p^\mu(\tilde{\xi})$ with arc capacity $\tilde{u}^\mu(x_j, x_i, \vartheta, \theta) = \tilde{\xi}(x_i, x_j, \theta, \vartheta) - \tilde{l}(x_i, x_j, \theta, \vartheta)$, transit time $\tau^\mu(x_j, x_i, \vartheta, \theta) = -\tau(x_i, x_j, \theta, \vartheta)$ and transmission cost $\tilde{c}^\mu(x_j, x_i, \vartheta, \theta) = -\tilde{c}(x_i, x_j, \theta, \vartheta)$.

If the maximum flow of the minimum cost equals the sum of the lower flow bounds wasn't found in the time-expanded graph without lower flow bounds, than there is no feasible flow in the time-expanded fuzzy graph without lower flow bounds and the task has no solution.

Based on the formulated rules and definitions represent method of solving this task.

Algorithm of the minimum cost flow finding in the dynamic network with fuzzy transit nonzero lower, upper flow bounds and transmission costs.

Let us represent formal method of the minimum cost flow finding in the dynamic network with fuzzy transit nonzero lower, upper flow bounds and transmission costs.

Step 1. Go to the time-expanded fuzzy static graph $\widetilde{G}_p = (X_p, \widetilde{A}_p)$ from the given fuzzy dynamic graph $\widetilde{G} = (X, \widetilde{A})$ expanding initial dynamic graph for the given number of time intervals making separate copy of each node $x_i \in X$ at each time period $\theta \in T$ according to the *rule 3.10*.

Step 2. Determine, if the task has a solution.

2.1. If $\tilde{\rho}(p) \geq \sum_{\tilde{l}(x_i, x_j, \vartheta, \theta) \neq \tilde{0}} \tilde{l}(x_i, x_j, \theta, \vartheta)$, turn to the **step 3**.

2.2. If $\tilde{\rho}(p) < \sum_{\tilde{l}(x_i, x_j, \vartheta, \theta) \neq \tilde{0}} \tilde{l}(x_i, x_j, \theta, \vartheta)$, the task has no solution, **the end**.

Step 3. Determine, if the time-expanded fuzzy graph $\widetilde{G}_p$ corresponding to the initial dynamic graph $\widetilde{G}$ has a feasible flow. Turn to the graph $\widetilde{G}_p^*$ according to the *rule 3.11*.

Step 4. Build a fuzzy residual network $\widetilde{G}_p^{*\mu} = (X_p^{*\mu}, \widetilde{A}_p^{*\mu})$ by the fuzzy network $\widetilde{G}_p^*$ according to the Definition 3.6.

Step 5. Search the augmenting minimum cost path $\widetilde{P}_p^{*\mu}$ from s^* to t^* in the $\widetilde{G}_p^{*\mu}$ according to the Ford's algorithm.

5.1. Go to the **step 6** if the augmenting path $\widetilde{P}_p^{*\mu}$ is found.

5.2. If the path is failed to find, the flow value $\tilde{\phi}^* < \sum_{\tilde{l}(x_i, x_j, \theta, \vartheta) \neq \tilde{0}} \tilde{l}(x_i, x_j, \theta, \vartheta)$, which is the maximum flow in $\widetilde{G}_p^*$, i.e. it is impossible to pass any additional flow unit, but not all the artificial arcs are saturated. Therefore, the time-expanded graph $\widetilde{G}_p$ has no feasible flow as the initial dynamic fuzzy graph $\widetilde{G}$ and the task has no solution. Exit.

Step 6. Pass $\tilde{\delta}_p^{*\mu} = \min[\tilde{u}(\widetilde{P}_p^{*\mu})]$, $\tilde{u}(\widetilde{P}_p^{*\mu}) = \min[\tilde{u}^{*\mu}(x_i, x_j, \theta, \vartheta)]$, $(x_i, \theta), (x_j, \vartheta) \in \widetilde{P}_p^{*\mu}$ flow units along the minimum cost path $\widetilde{P}_p^{*\mu}$.

Step 7. Update the fuzzy flow values in the graph $\widetilde{G}_p^*$: replace the fuzzy flow $\tilde{\xi}^*(x_j, x_i, \theta, \vartheta)$ along the corresponding arcs going from (x_j^*, θ) to (x_i^*, ϑ) from $\widetilde{G}_p^*$ by $\tilde{\xi}^*(x_j, x_i, \theta, \vartheta) - \tilde{\delta}_p^{*\mu}$ for arcs connecting node-time pair $(x_i^{*\mu}, \vartheta)$ with $(x_j^{*\mu}, \theta)$ in $\widetilde{G}_p^{*\mu}$, such as $((x_i^{*\mu}, \vartheta), (x_j^{*\mu}, \theta)) \notin \widetilde{A}_p^*$, $((x_i^{*\mu}, \vartheta), (x_j^{*\mu}, \theta)) \in \widetilde{A}_p^{*\mu}$ and replace the fuzzy flow $\tilde{\xi}^*(x_i, x_j, \theta, \vartheta)$ along the arcs going from (x_i^*, θ) to (x_j^*, ϑ) from $\widetilde{G}_p^*$ by $\tilde{\xi}^*(x_i, x_j, \theta, \vartheta) + \tilde{\delta}_p^{*\mu}$ for arcs connecting node-time pair $(x_i^{*\mu}, \theta)$ with $(x_j^{*\mu}, \vartheta)$ in $\widetilde{G}_p^{*\mu}$, such as, $((x_i^{*\mu}, \theta), (x_j^{*\mu}, \vartheta)) \in \widetilde{A}_p^*$, $((x_i^{*\mu}, \theta), (x_j^{*\mu}, \vartheta)) \in \widetilde{A}_p^{*\mu}$. Replace $\tilde{\xi}^*(x_i, x_j, \theta, \vartheta)$ by $\tilde{\xi}^*(x_i, x_j, \theta, \vartheta) + \tilde{\delta}_p^{*\mu} \widetilde{P}_p^{*\mu}$ and turn to the step 8.

Step 8. Compare $\tilde{\xi}^*(x_i, x_j, \theta, \vartheta) + \tilde{\delta}_p^{*\mu} \widetilde{P}_p^{*\mu}$ and $\sum_{\tilde{l}(x_i, x_j, \theta, \vartheta) \neq \tilde{0}} \tilde{l}(x_i, x_j, \theta, \vartheta)$:

8.1. If the flow value $\tilde{\xi}^*(x_i, x_j, \theta, \vartheta) + \tilde{\delta}_p^{*\mu} \widetilde{P}_p^{*\mu}$ of the minimum cost of the value $\tilde{\sigma}^*$ is less than $\sum_{\tilde{l}(x_j, x_i, \vartheta, \theta) \neq \tilde{0}} \tilde{l}(x_i, x_j, \theta, \vartheta)$ and less than given floe value, i.e. $\tilde{\sigma}^* \sum_{\tilde{l}(x_i, x_j, \theta, \vartheta) \neq \tilde{0}} \tilde{l}(x_i, x_j, \theta, \vartheta) \leq \tilde{p}(p)$ and not all artificial arcs become saturated, go to the **step 4**.

8.2. If the flow value $\tilde{\xi}^*(x_i, x_j, \theta, \vartheta) + \tilde{\delta}_p^{*\mu} \widetilde{P}_p^{*\mu}$ of the minimum cost of the value $\tilde{\sigma}^*$ is equal to $\sum_{\tilde{l}(x_j, x_i, \theta, \vartheta) \neq \tilde{0}} \tilde{l}(x_i, x_j, \vartheta, \theta)$ and doesn't exceed $\tilde{p}(p)$, i.e. $\tilde{\sigma}^* \sum_{\tilde{l}(x_i, x_j, \theta, \vartheta) \neq \tilde{0}} \tilde{l}(x_i, x_j, \theta, \vartheta) \leq \tilde{p}(p)$ and all arcs from the artificial source to the artificial sink become saturated, then the value $\tilde{\xi}^*(x_i, x_j, \theta, \vartheta) + \tilde{\delta}_p^{*\mu} \widetilde{P}_p^{*\mu}$ is required value of maximum flow $\tilde{\sigma}^*$ of the minimum cost $\tilde{c}(\tilde{\xi}^*(x_i, x_j, \theta, \vartheta) + \tilde{\delta}_p^{*\mu} \widetilde{P}_p^{*\mu})$. In this case the total flow along the artificial arcs connecting the node-time pairs $(t, \forall \theta \in T)$ with $(s, \forall \theta \in T)$, which is equal to $\sum_{\theta=0}^{p} \tilde{\xi}^*(t, s, \forall \theta \in T, \forall \theta \in T)$ in $\widetilde{G}_p^*$ determines the feasible flow in the time-expanded graph $\widetilde{G}_p$ with the flow value $\sum_{\theta=0}^{p} \tilde{\xi}^*(t, s, \forall \theta \in T, \forall \theta \in T) = \tilde{\sigma}$ of the minimum cost. Turn to the graph $\widetilde{G}_p$ from the graph $\widetilde{G}_p^*$ according to the *rule 3.5*. The network $\widetilde{G}_p(\tilde{\xi})$ is obtained. Determine if the flow is optimal.

8.2.1. If the flow in $\widetilde{G}_p(\tilde{\xi})$ is equal to $\tilde{\rho}(p)$ of the cost $\sum_{\theta=0}^{p} \sum_{(x_i, x_j) \in A} \tilde{c}_{ij}(\theta) \tilde{\xi}_{ij}(\theta)$, we found the required minimum cost flow, exit.

8.2.2. If the flow in $\widetilde{G}_p(\tilde{\xi})$ is less than $\tilde{\rho}(p)$ of the cost $\sum_{\theta=0}^{p} \sum_{(x_i, x_j) \in A} \tilde{c}_{ij}(\theta) \tilde{\xi}_{ij}(\theta)$, we obtain the network $\widetilde{G}(\tilde{\xi})$ with the feasible flow. Turn to the **step 9**.

Step 9. Construct the residual network $\widetilde{G}_p^\mu(\tilde{\xi})$ taking into account the feasible flow vector $\tilde{\xi} = (\tilde{\xi}(x_i, x_j, \theta, \vartheta))$ in $\widetilde{G}_p(\tilde{\xi})$ adding the artificial source and sink and the arcs with infinite arc capacity, connecting s' with true sources and t' with true sinks according to the *rule 3.12*.

Step 10. Define the shortest path $\widetilde{P}_p^\mu$ according to the breadth-first-search from s' в t' in the constructed residual network $\widetilde{G}_p^\mu(\tilde{\xi})$.

10.1. Go to the **step 11** if the augmenting path $\widetilde{P}_p^\mu$ is found.

10.2. If the path is failed to find, then given flow value exceeds the maximum flow value in $\widetilde{G}$, i.e. $\tilde{\rho}(p) > \tilde{v}(p)$ and the task has no solution, the end.

Step 11. Pass the flow value, $\tilde{\delta}_p^\mu = \min[\tilde{u}(\widetilde{P}_p^\mu)]$, $\tilde{u}(\widetilde{P}_p^\mu) = \min[\tilde{u}^\mu(x_i, x_j, \theta, \vartheta)]$, $(x_i, \theta), (x_j, \vartheta) \in \widetilde{P}_p^\mu$ along the found path.

Step 12. Update the flow values in the graph $\widetilde{G}_p(\tilde{\xi})$: replace the flow $\tilde{\xi}(x_j, x_i, \theta, \vartheta)$ by $\tilde{\xi}(x_j, x_i, \theta, \vartheta) - \tilde{\delta}_p^\mu$ along the corresponding arcs, going from (x_j, θ) to (x_i, ϑ) from $\widetilde{G}_p(\tilde{\xi})$ for arcs, connecting node-time pair (x_i^μ, ϑ) with (x_j^μ, θ) in $\widetilde{G}_p^\mu(\tilde{\xi})$, such as, $((x_i^\mu, \vartheta), (x_j^\mu, \theta)) \notin \widetilde{A}_p$, $((x_i^\mu, \vartheta), (x_j^\mu, \theta)) \in \widetilde{A}_p^\mu$, and replace the flow $\tilde{\xi}(x_i, x_j, \theta, \vartheta)$ by $\tilde{\xi}(x_i, x_j, \theta, \vartheta) + \tilde{\delta}_p^\mu$ along the corresponding arcs, going from (x_i, θ) to (x_j, ϑ) from $\widetilde{G}_p(\tilde{\xi})$ for arcs, connecting node-time pair (x_i^μ, θ) with (x_j^μ, ϑ) in $\widetilde{G}_p^\mu(\tilde{\xi})$, such as, $((x_i^\mu, \theta), (x_j^\mu, \vartheta)) \in \widetilde{A}_p$, $((x_i^\mu, \theta), (x_j^\mu, \vartheta)) \in \widetilde{A}_p^\mu$. Replace the flow value in $\widetilde{G}_p$: $\tilde{\xi}(x_i, x_j, \theta, \vartheta)$ by $\tilde{\xi}(x_i, x_j, \theta, \vartheta) + \tilde{\delta}_p^\mu \widetilde{P}_p^\mu$ and turn to the **step 13**, starting from the new flow value.

Step 13. Compare the flow value $\tilde{\xi}(x_i, x_j, \theta, \vartheta) + \tilde{\delta}_p^\mu \widetilde{P}_p^\mu$ and $\tilde{\rho}(p)$:

13.1. If the flow value $\tilde{\xi}(x_i, x_j, \theta, \vartheta) + \tilde{\delta}_p^\mu \widetilde{P}_p^\mu$ of the minimum cost $\tilde{c}(\tilde{\xi}(x_i, x_j, \theta, \vartheta) + \tilde{\delta}_p^\mu \widetilde{P}_p^\mu)$ is less than $\tilde{\rho}(p)$, replace $\tilde{\xi}(x_i, x_j, \theta, \vartheta)$ by $\tilde{\xi}(x_i, x_j, \theta, \vartheta) + \tilde{\delta}_p^\mu \widetilde{P}_p^\mu$ and turn to the **step 9**.

13.2. If the flow value $\tilde{\xi}(x_i, x_j, \theta, \vartheta) + \tilde{\delta}_p^\mu \widetilde{P}_p^\mu$ of the minimum cost $\tilde{c}(\tilde{\xi}(x_i, x_j, \theta, \vartheta) + \tilde{\delta}_p^\mu \widetilde{P}_p^\mu)$ equals $\tilde{\rho}(p)$, therefore, the given flow value of the minimum cost is found, turn to the **step 14**.

13.3. If the flow value $\tilde{\xi}(x_i, x_j, \theta, \vartheta) + \tilde{\delta}_p^\mu \widetilde{P}_p^\mu = \tilde{h}(p)$, where $\tilde{h}(p) > \tilde{\rho}(p)$ of the minimum cost $\tilde{c}(\tilde{\xi}(x_i, x_j, \theta, \vartheta) + \tilde{\delta}_p^\mu \widetilde{P}_p^\mu)$, then required flow of the minimum cost will be $\tilde{\xi}(x_i, x_j, \theta, \vartheta) + (\tilde{\delta}_p^\mu - \tilde{h}(p) + \tilde{\rho}(p))\tilde{p}_p^\mu$ of the minimum cost $\tilde{c}(\tilde{\xi}(x_i, x_j, \theta, \vartheta) + (\tilde{\delta}_p^\mu - \tilde{h}(p) + \tilde{\rho}(p))\tilde{p}_p^\mu)$, turn to the **step 14**.

Step 14. If the flow is $\tilde{\xi}(x_i, x_j, \theta, \vartheta) + \tilde{\delta}_p^\mu \times \tilde{P}_p^\mu = \tilde{\rho}(p)$ in the graph $\tilde{G}_p(\tilde{\xi})$ of the minimum cost $\tilde{c}(\tilde{\xi}(x_i, x_j, \theta, \vartheta) + \tilde{\delta}_p^\mu \times \tilde{P}_p^\mu)$ from the artificial source s' to the artificial sink t', defined by the set of paths $\tilde{P}_p^\mu$, turn to the initial dynamic graph $\tilde{G}$ such a way: delete artificial nodes s', t' and arcs, connecting them with other nodes. Thus, the flow of the value $\tilde{\rho}(p)$ of the minimum cost is obtained in the initial dynamic graph, which is equivalent to the flow from the sources (the initial node expanded to the p intervals) to the sinks (terminal node expanded to the p intervals) in $\tilde{G}_p$ after deleting artificial nodes and each path, connecting nodes (s, ς) and $(t, \psi = \varsigma + \tau_{st}(\varsigma)), \psi \in T$, with the flow $\tilde{\xi}(s, t, \theta, \varsigma)$ of the cost $\tilde{c}(\tilde{\xi}(s, t, \theta, \varsigma))$, coming along, corresponds to the flow $\tilde{\xi}_{st}(\theta)$ of the cost $\tilde{c}(\tilde{\xi}_{st}(\theta))$.

Consider example, illustrating the algorithm of the minimum cost flow finding with nonzero lower flow bounds in dynamic network in fuzzy conditions.

Numerical example 9

Let the network, which is the part of the railway network, is represented in the form of the fuzzy directed graph, obtained from «ObjectLand» (Fig. 3.40).

It is necessary to the find the minimum transmission cost of $3\tilde{8}$ flow units in the dynamic graph with transit lower, upper flow bounds for 3 time periods and и represent the result in the form of the fuzzy triangular number, if there are basic values of arc capacities, transmission costs set as fuzzy triangular numbers, as shown in Figs. 2.27–2.304. Fuzzy lower, upper flow bounds, transmission costs and time parameters are transit and represented in Tables 3.9, 3.10, 3.11 and 3.12.

Step 1. Turn to the time-expanded for p intervals fuzzy static graph $\tilde{G}_p$ from given fuzzy dynamic graph $\tilde{G}$ according to the *rule 3.10*, as shown in Fig. 3.41.

Fig. 3.40 Initial dynamic graph $\tilde{G}$

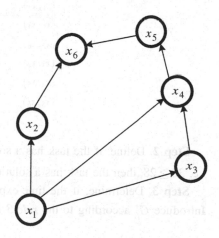

Table 3.9 Fuzzy upper flow bounds $\tilde{u}_{ij}$, depend on the flow departure time θ

Arcs of the graph	Fuzzy upper flow bounds $\tilde{u}_{ij}$ at the time θ, units.			
	0	1	2	3
(x_1, x_2)	$\widetilde{10}$	$\widetilde{18}$	$\tilde{8}$	$\widetilde{10}$
(x_1, x_3)	$\widetilde{10}$	$\widetilde{15}$	$\widetilde{18}$	$\widetilde{18}$
(x_1, x_4)	$\widetilde{20}$	$\widetilde{20}$	$\widetilde{25}$	$\widetilde{18}$
(x_2, x_6)	$\widetilde{30}$	$\widetilde{25}$	$\widetilde{40}$	$\widetilde{30}$
(x_3, x_4)	$\widetilde{25}$	$\widetilde{18}$	$\widetilde{20}$	$\widetilde{20}$
(x_4, x_5)	$\widetilde{30}$	$\widetilde{30}$	$\widetilde{25}$	$\widetilde{25}$
(x_5, x_6)	$\widetilde{30}$	$\widetilde{30}$	$\widetilde{45}$	$\widetilde{25}$

Table 3.10 Fuzzy lower flow bounds $\tilde{l}_{ij}$, depend on the flow departure time θ

Arcs of the graph	Fuzzy lower flow bounds $\tilde{u}_{ij}$ at the time θ, units.			
	0	1	2	3
(x_1, x_2)	$\tilde{0}$	$\widetilde{10}$	$\tilde{0}$	$\tilde{0}$
(x_1, x_3)	$\tilde{0}$	$\tilde{0}$	$\tilde{0}$	$\tilde{0}$
(x_1, x_4)	$\widetilde{18}$	$\tilde{0}$	$\tilde{0}$	$\tilde{0}$
(x_2, x_6)	$\tilde{0}$	$\tilde{0}$	$\tilde{0}$	$\tilde{0}$
(x_3, x_4)	$\tilde{0}$	$\tilde{0}$	$\tilde{0}$	$\tilde{0}$
(x_4, x_5)	$\tilde{0}$	$\tilde{0}$	$\tilde{0}$	$\tilde{0}$
(x_5, x_6)	$\tilde{0}$	$\tilde{0}$	$\tilde{0}$	$\tilde{0}$

Table 3.11 Fuzzy transmission costs $\tilde{c}_{ij}$, depend on the flow departure time θ

Arcs of the graph	Fuzzy transmission costs $\tilde{c}_{ij}$ at the time θ, units.			
	0	1	2	3
(x_1, x_2)	$\widetilde{30}$	$\widetilde{60}$	$\widetilde{60}$	$\widetilde{30}$
(x_1, x_3)	$\widetilde{75}$	$\widetilde{50}$	$\widetilde{18}$	$\widetilde{18}$
(x_1, x_4)	$\widetilde{70}$	$\widetilde{30}$	$\widetilde{20}$	$\widetilde{18}$
(x_2, x_6)	$\widetilde{30}$	$\widetilde{25}$	$\widetilde{50}$	$\widetilde{30}$
(x_3, x_4)	$\widetilde{25}$	$\widetilde{30}$	$\widetilde{80}$	$\widetilde{20}$
(x_4, x_5)	$\widetilde{30}$	$\widetilde{100}$	$\widetilde{80}$	$\widetilde{25}$
(x_5, x_6)	$\widetilde{30}$	$\widetilde{30}$	$\widetilde{80}$	$\widetilde{25}$

Step 2. Define, if the task has a solution. As $\tilde{\rho}(p) \geq \sum_{\tilde{l}(x_i, x_j, \theta) \neq \tilde{0}} \tilde{l}(x_i, x_j, \theta, \vartheta)$, i.e. $3\tilde{8} \geq 2\tilde{8}$, then the task has a solution, turn to the **step 3**.

Step 3. Determine, if the time-expanded fuzzy graph $\widetilde{G}_p$ has a feasible flow. Introduce $\widetilde{G}_p^*$ according to the *rule 3.11*, as shown in Fig. 3.42.

Arcs of the graph	Transit time parameters τ_{ij} at the time θ, time units.			
	0	1	2	3
(x_1, x_2)	5	1	3	2
(x_1, x_3)	1	3	2	1
(x_1, x_4)	1	3	3	3
(x_2, x_6)	4	4	1	2
(x_3, x_4)	5	1	2	3
(x_4, x_5)	4	1	1	1
(x_5, x_6)	4	4	1	1

Table 3.12 Transit time parameters τ_{ij}, depend on the flow departure time θ

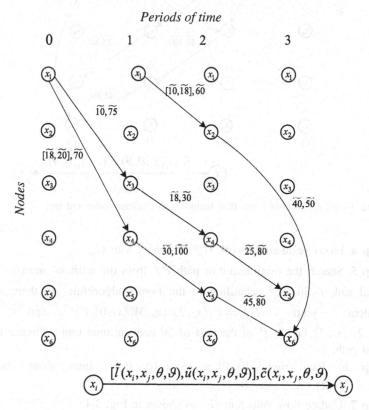

Fig. 3.41 Graph $\widetilde{G}_p$ «time-expanded» graph $\widetilde{G}$

Connectors of the same form connect the corresponding pair of nodes. Connectors, which have the same shape (for example, ▲) link the corresponding pair of nodes in Fig. 3.42. Every arc, going from x_6 for all time periods to the nodes x_1 for all time periods have infinite upper flow bounds.

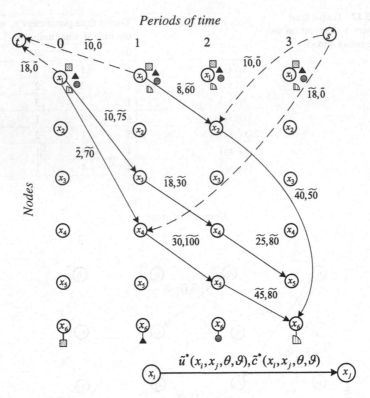

Fig. 3.42 Graph $\widetilde{G}_p^*$ without lower flow bounds with artificial nodes and arcs

Step 4. Fuzzy residual network $\widetilde{G}_p^{*\mu}$ coincides with $\widetilde{G}_p^*$.

Step 5. Search the minimum cost path $\widetilde{P}_p^{*\mu}$ from the artificial source s^* to the artificial sink t^* in $\widetilde{G}_p^{*\mu}$ according to the Ford's algorithm. As there are two equivalent path $\widetilde{P}_p^{*\mu} = s^*, (x_2, 2), (x_6, 3), (x_1, 0), t^*$ and $\widetilde{P}_p^{*\mu} = s^*, (x_2, 2), (x_6, 3), (x_1, 1), t^*$ of the cost of $5\tilde{0}$ conventional units. Choose the first shortest path.

Step 6. Pass min $[1\tilde{0}, 4\tilde{0}, \infty, 1\tilde{8}] = 10$ flow units along the path $\widetilde{P}_1^{*\mu} = s^*, (x_2, 2), (x_6, 3), (x_1, 0), t^*$.

Step 7. Update flow values in $\widetilde{G}_p^*$, as shown in Fig. 3.43.

The flow $\widetilde{\xi}^*(x_i, x_j, \theta, \vartheta) = \tilde{0}$ turns to $\widetilde{\xi}^*(x_i, x_j, \theta, \vartheta) = \tilde{0} + 1\tilde{0} = 1\tilde{0}$ flow units.

Step 8.1. As the flow value of 10 units is less than $\sum_{\tilde{l}(x_j, x_i, \vartheta, \theta) \neq \tilde{0}} \tilde{l}(x_i, x_j, \theta, \vartheta) = 2\tilde{8}$ and less than given flow value 38 units, i.e. $\tilde{\sigma}^* < \sum_{\tilde{l}(x_i, x_j, \theta, \vartheta) \neq \tilde{0}} \tilde{l}(x_i, x_j, \theta, \vartheta) \leq \tilde{\rho}(p)$, turn to the construction of the residual network with new flow distribution, as shown in Fig. 3.44.

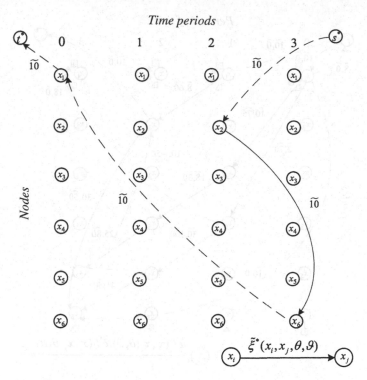

Fig. 3.43 Graph $\widetilde{G}_p^*$ with the flow of $1\tilde{0}$ units

Step 5. Search the second shortest path of the minimum cost $\widetilde{P}_2^{*\mu}$ in the residual network $\widetilde{G}_p^{*\mu}$, presented in Fig. 3.44. As two equivalent paths were obtained on this step $\widetilde{P}_2^{*\mu} = s^*, (x_4, 1), (x_5, 2), (x_6, 3)(x_1, 0), t^*$ and $\widetilde{P}_2^{*\mu} = s^*, (x_4, 1), (x_5, 2), (x_6, 3)$ $(x_1, 1), t^*$ of the cost 180 conventional units. Choose the first path.

Step 6. Pass $\min(1\tilde{8}, 3\tilde{0}, 4\tilde{5}, \infty, \tilde{8}) = \tilde{8}$, i.e. $\tilde{8}$ flow units along the path $\widetilde{P}_2^{*\mu} = s^*, (x_4, 1), (x_5, 2), (x_6, 3)(x_1, 0), t^*$.

Step 7. Update flow values in $\widetilde{G}_p^*$ and construct graph with the new flow value, as shown in Fig. 3.45.

Flow $\tilde{\xi}^*(x_i, x_j, \theta, \vartheta) = 1\tilde{0}$ turns to $\tilde{\xi}^*(x_i, x_j, \theta, \vartheta) = 1\tilde{0} + \tilde{8} = 1\tilde{8}$ flow units.

Step 8.1. As the flow value of $1\tilde{8}$ flow units is less than $\sum_{\tilde{l}(x_j, x_i, \vartheta, \theta) \neq \tilde{0}} \tilde{l}(x_i, x_j, \theta, \vartheta) = 2\tilde{8}$ and less than the given flow value of $3\tilde{8}$ units, i.e. $\tilde{\sigma}^* < \sum_{\tilde{l}(x_i, x_j, \theta, \vartheta) \neq \tilde{0}} \tilde{l}(x_i, x_j, \theta, \vartheta) \leq \tilde{p}(p)$, turn to the construction of the residual network with new flow distribution, as shown in Fig. 3.46.

Step 5. Find the third path of the minimum cost $\widetilde{P}_3^{*\mu}$ in the residual network $\widetilde{G}_p^{*\mu}$, presented in Fig. 3.46. The path $\widetilde{P}_3^{*\mu} = s^*, (x_4, 1), (x_5, 2), (x_6, 3), (x_1, 1), t^*$ is obtained of the cost $18\tilde{0}$ conventional units.

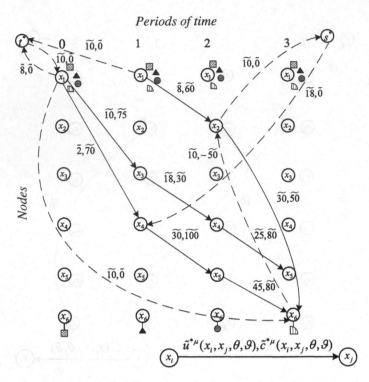

Fig. 3.44 Fuzzy residual network $\widetilde{G}_p^{*\mu}$ for the graph in Fig. 3.43

Step 6. Pass $\min(1\tilde{0}, 2\tilde{2}, 3\tilde{7}, \infty, 1\tilde{0}) = \tilde{8}$, i.e. $1\tilde{0}$ flow units along the path $\widetilde{P}_3^{*\mu} = s^*, (x_4, 1), (x_5, 2), (x_6, 3)(x_1, 1), t^*$.

Step 7. Update flow values in $\widetilde{G}_p^*$ and construct the graph with the new flow distribution, as shown in Fig. 3.47.

Step 8.3. As the flow value of $2\tilde{8}$ is less than $\sum_{\tilde{l}(x_j, x_i, \vartheta, \theta) \neq \tilde{0}} \tilde{l}(x_i, x_j, \theta, \vartheta) = 2\tilde{8}$ and less than given flow value of $3\tilde{8}$ units, i.e. $\tilde{\sigma}^* \sum_{\tilde{l}(x_i, x_j, \theta, \vartheta) \neq \tilde{0}} \tilde{l}(x_i, x_j, \theta, \vartheta) \leq \tilde{\rho}$, maximum flow in the time-expanded graph $\widetilde{G}_p^*$ of the minimum cost is found. This flow is equal to the sum of the lower flow bounds, passing along the artificial arcs. Therefore, feasible flow in $\widetilde{G}_p$ exists, equals the total flow, passing along the reverse arcs, connecting the nodes $(x_6, \forall \theta \in T)$ and $(x_1, \forall \theta \in T)$ at all time periods, i.e. $1\tilde{0} + \tilde{8} + 1\tilde{0} = 2\tilde{8}$ flow units. Build the network $\widetilde{G}_p(\tilde{\xi})$ with the flow for the minimum cost flow finding in $\widetilde{G}_p$, deleting artificial nodes and arcs and taking into account, that the feasible flow vector $\tilde{\xi} = (\tilde{\xi}(x_i, x_j, \theta, \vartheta))$ of the value $\tilde{\sigma}$, is defined as: $\tilde{\xi}(x_i, x_j, \theta, \vartheta) = \tilde{\xi}^*(x_i, x_j, \theta, \vartheta) + l(x_i, x_j, \theta, \vartheta)$, where $\tilde{\xi}^*(x_i, x_j, \theta, \vartheta)$—flows, passing along the arcs of the graph $\widetilde{G}_p^*$ after deleting artificial nodes and arcs, as shown in Fig. 3.48.

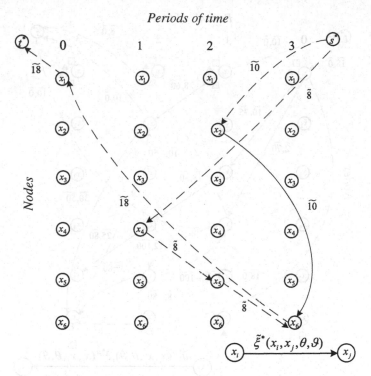

Fig. 3.45 Graph $\widetilde{G}_p^*$ with the flow $\widetilde{18}$ units

The flow $\tilde{\xi}^*(x_i, x_j, \theta, \vartheta) = \widetilde{18}$ turns to $\tilde{\xi}^*(x_i, x_j, \theta, \vartheta) = \widetilde{18} + \widetilde{10} = \widetilde{28}$ flow units.

Step 9. Introduce artificial source and sink, connecting them by arcs with infinite arc capacities with true sources and sinks, build residual network $\widetilde{G}_p^\mu(\tilde{\xi})$ for the graph in Fig. 3.48, as shown in Fig. 3.49.

Step 10.1. Find the first path of the minimum cost $\widetilde{P}_1^\mu$ in $\widetilde{G}_p^\mu(\tilde{\xi})$:
$\widetilde{P}_1^\mu = s', (x_1, 1), (x_2, 2), (x_6, 3), t'.$

Step 11. Pass $\min(\infty, \tilde{8}, \widetilde{30}, \infty) = \tilde{8}$, i.e. $\tilde{8}$ flow units along the path $\widetilde{P}_1^\mu = s', (x_1, 1), (x_2, 2), (x_6, 3), t'.$

Step 12. Update flow values in $\widetilde{G}_p(\tilde{\xi})$ and construct a graph with the new flow value, as shown in Fig. 3.50.

The flow $\tilde{\xi}(x_i, x_j, \theta, \vartheta) = \widetilde{28}$ turns to $\tilde{\xi}^*(x_i, x_j, \theta, \vartheta) = \widetilde{28} + \tilde{8} = \widetilde{36}$ flow units.

Step 13.1. Compare flow values $\tilde{\xi}(x_i, x_j, \theta, \vartheta) + \tilde{\delta}_p^\mu \widetilde{P}_p^\mu$ and $\tilde{p}(p)$. As $\tilde{\xi}(x_i, x_j, \theta, \vartheta) + \tilde{\delta}_p^\mu \widetilde{P}_p^\mu = \widetilde{36}$, where $\widetilde{36} < \widetilde{38}$, then turn to the **step 9**.

Step 9. Build a network $\widetilde{G}_p^\mu(\tilde{\xi})$ for the graph in Fig. 3.50, as shown in Fig. 3.51.

Step 10.1. Find the second shortest path of the minimum cost $\widetilde{P}_2^\mu$ in $\widetilde{G}_p^\mu(\tilde{\xi})$:
$\widetilde{P}_1^\mu = s', (x_1, 0), (x_3, 1), (x_4, 2), (x_6, 3), t'.$

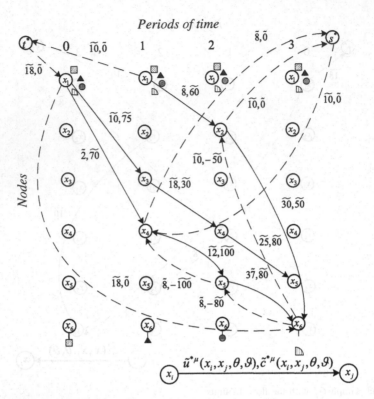

Fig. 3.46 Fuzzy residual network $\widetilde{G}_p^{*\mu}$ for the graph in Fig. 3.45

Step 11. Pass $\min(\infty, \widetilde{10}, \widetilde{18}, \widetilde{25}, \infty) = \widetilde{10}$, i.e. $\widetilde{10}$ flow units along the path $\widetilde{P}_1^\mu = s', (x_1, 0), (x_3, 1), (x_4, 2), (x_6, 3), t'$.

Step 12. Update flow values in $\widetilde{G}_p(\widetilde{\xi})$ and construct the graph with the new flow value, as shown in Fig. 3.52.

Step 13.2. Compare flow values $\widetilde{\xi}(x_i, x_j, \theta, \vartheta) + \widetilde{\delta}_p^\mu \widetilde{P}_p^\mu$ and $\widetilde{\rho}(p)$. As $\widetilde{\xi}(x_i, x_j, \theta, \vartheta) + \widetilde{\delta}_p^\mu \widetilde{P}_p^\mu = \widetilde{38}$, turn to the **step 14**.

Step 14. Construct a graph $\widetilde{G}_p$ with the flow $\widetilde{38}$ units, as shown in Fig. 3.53.

Transmission cost of $\widetilde{38}$ flow units is $\widetilde{20} \cdot \widetilde{70} + \widetilde{18} \cdot \widetilde{60} + \widetilde{18} \cdot \widetilde{50} + \widetilde{20} \cdot \widetilde{100} + \widetilde{20} \cdot \widetilde{80} = 6980$ conventional units.

Turning to the dynamic graph $\widetilde{G}$ from the «time-expanded» static graph $\widetilde{G}_p$, it can be concluded, that dynamic flow of the minimum cost for 3 periods is equal to the flow leaving the node-time pairs $(x_1, 0)$ and $(x_1, 1)$ and entering the node-time pair $(x_6, 3)$ and is equal to $\widetilde{38}$ flow units, which are defined by two paths: path $x_1 \to x_2 \to x_6$, which the flow departures along at the moment $\theta = 1$ and arriving at the sink at the moment $\theta = 3$, path $x_1 \to x_4 \to x_5 \to x_6$, which the flow departures along the $\theta = 0$ and arriving at the sink at the moment $\theta = 3$.

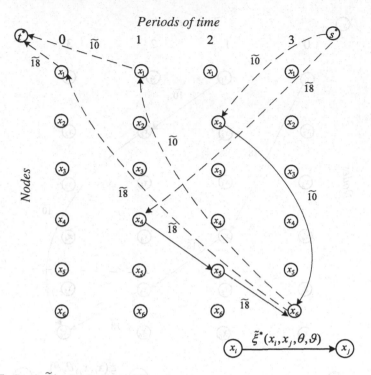

Fig. 3.47 Graph $\widetilde{G}_p^*$ with the flow of 28 units

Define borders of uncertainty of the fuzzy number $3\widetilde{8}$, corresponding to the given flow in $\widetilde{G}$. Obtained result is between two basic neighboring values of fuzzy arc capacities: $3\widetilde{1}$ with the left deviation $l_1^L = 8$, right deviation—$l_1^R = 7$ and $4\widetilde{4}$ with the left deviation $l_2^L = 9$, right deviation—$l_2^L = 10$. According to Eq. (2.4) it is obtained:

$$l^L = \frac{(a_2 - a')}{(a_2 - a_1)} \times l_1^L + \left(1 - \frac{(a_2 - a')}{(a_2 - a_1)}\right) \times l_2^L$$
$$= \frac{(44 - 38)}{(44 - 31)} \times 8 + \left(1 - \frac{(44 - 38)}{(44 - 31)}\right) \times 9$$
$$= 8.56 \approx 9,$$

$$l^R = \frac{(a_2 - a')}{(a_2 - a_1)} \times l_1^R + \left(1 - \frac{(a_2 - a')}{(a_2 - a_1)}\right) \times l_2^R$$
$$= \frac{(44 - 38)}{(44 - 31)} \times 7 + \left(1 - \frac{(44 - 38)}{(44 - 31)}\right) \times 10$$
$$= 8.63 \approx 9.$$

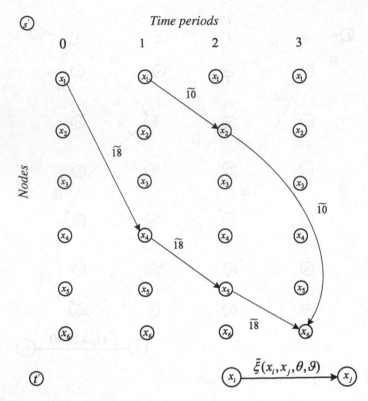

Fig. 3.48 Graph $\widetilde{G}_p(\widetilde{\xi})$ with the feasible flow $\widetilde{\xi} = (\widetilde{\xi}(x_i, x_j, \theta, \vartheta))$

Therefore, we can represent the value of the given flow in the fuzzy number of the triangular form (38,9,9), as shown in Fig. 3.55.

According to Fig. 3.55 the flow of (38,9,9) units with the degree of confidence 0,7 will be in the interval [35,41], but anyway the minimum cost flow will be no less than 29 units and no more than 47 units.

Present the minimum transmission cost of (38,9,9) flow units equals $698\widetilde{0}$ conventional units in the form of the fuzzy triangular number.

Let us define borders of uncertainty of the obtained fuzzy number of $698\widetilde{0}$ conventional units. The obtained result is between two neighboring basic values of fuzzy transportation costs: $662\widetilde{0}$ with the left deviation $l_2^L = 620$, right deviation— $l_2^R = 710$ and $754\widetilde{0}$ with the left deviation $l_2^L = 690$, right deviation—$l_2^R = 760$. According to (2.4) we obtain:

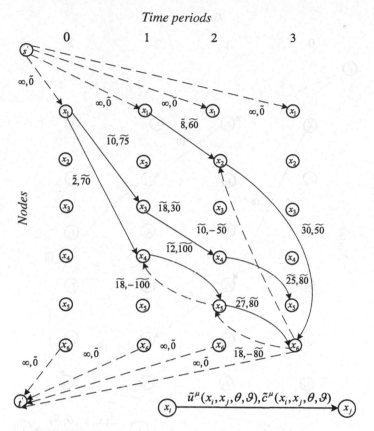

Fig. 3.49 Fuzzy residual network $\widetilde{G}_p^\mu(\widetilde{\xi})$ for the graph in Fig. 3.48

$$l^L = \frac{(a_2 - a')}{(a_2 - a_1)} \times l_1^L + \left(1 - \frac{(a_2 - a')}{(a_2 - a_1)}\right) \times l_2^L$$

$$= \frac{(7540 - 6980)}{(7540 - 6620)} \times 620 + \left(1 - \frac{(7540 - 6980)}{(7540 - 6620)}\right) \times 690$$

$$= 646.5 \approx 647,$$

$$l^R = \frac{(a_2 - a')}{(a_2 - a_1)} \times l_1^R + \left(1 - \frac{(a_2 - a')}{(a_2 - a_1)}\right) \times l_2^R$$

$$= \frac{(7540 - 6980)}{(7540 - 6620)} \times 710 + \left(1 - \frac{(7540 - 6980)}{(7540 - 6620)}\right) \times 760$$

$$= 728.57 \approx 729.$$

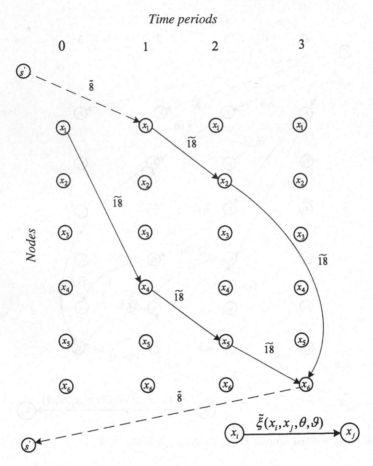

Fig. 3.50 Graph $\widetilde{G}_p(\widetilde{\xi})$ after finding the first path of minimum cost

Therefore, we can represent the minimum transportation flow cost in the form of the triangular fuzzy number of (6980,647,729) conventional units, as shown in Fig. 3.54.

According to Fig. 3.55 the minimum transportation cost of (38,9,9) flow units with the degree of confidence 0,8 will be within the interval [6851,7126] conventional unit, but anyway the minimum transportation cost of (38,9,9) flow units will be no less than 6333 and no more than 7709 conventional units.

Minimum cost maximum flow finding in the dynamic network with fuzzy transit nonzero lower, upper flow bounds and transmission costs.

Consider situation, when it is necessary to transport maximum flow from the initial point to terminal, taking into account instantaneous flow passing along the arcs of the network and minimum value of economical reliability of transportation,

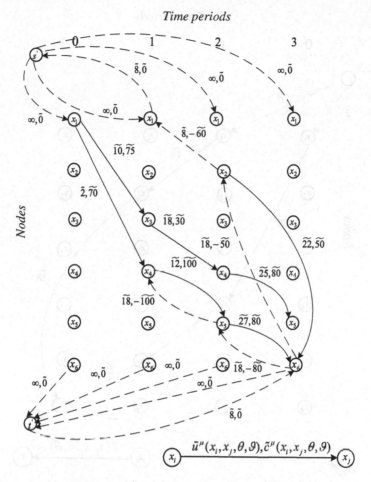

Fig. 3.51 Fuzzy residual network $\widetilde{G}_p^{\mu}(\tilde{\xi})$ for the graph in Fig. 3.50

expressed in lower flow bounds. Parameters of the network are set in the fuzzy form. The problem statement of this task can be represented as follows [16]:

$$Minimize \sum_{\theta=0}^{p} \sum_{(x_i,x_j)\in\widetilde{A}} \tilde{c}_{ij}(\theta)\tilde{\xi}_{ij}(\theta), \forall(x_i,x_j)\in\widetilde{A}, \quad \theta\in T, \qquad (3.34)$$

$$\sum_{\theta=0}^{p} \left(\sum_{x_j\in\Gamma(x_i)} \tilde{\xi}_{ij}(\theta) - \sum_{x_j\in\Gamma^{-1}(x_i)} \tilde{\xi}_{ij}(\theta-\tau_{ji}(\theta)) \right) = \tilde{v}(p), \quad x_i = s, \qquad (3.35)$$

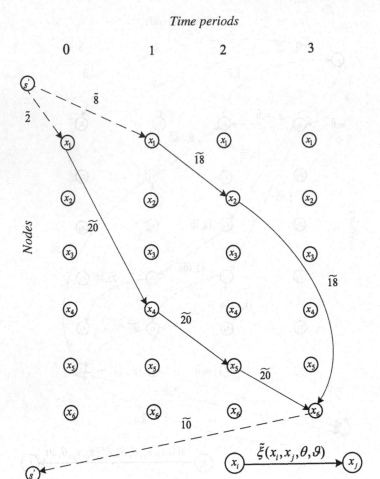

Fig. 3.52 Graph $\widetilde{G}_p(\widetilde{\xi})$ after finding the second path of the minimum cost

$$\sum_{\theta=0}^{p}\left(\sum_{x_j\in\Gamma(x_i)}\widetilde{\xi}_{ij}(\theta)-\sum_{x_j\in\Gamma^{-1}(x_i)}\widetilde{\xi}_{ji}(\theta-\tau_{ji}(\theta))\right)=\widetilde{0},\quad x_i\neq s,t;\ \theta\in T, \qquad (3.36)$$

$$\sum_{\theta=0}^{p}\left(\sum_{x_j\in\Gamma(x_i)}\widetilde{\xi}_{ij}(\theta)-\sum_{x_j\in\Gamma^{-1}(x_i)}\widetilde{\xi}_{ji}(\theta-\tau_{ji}(\theta))\right)=-\widetilde{v}(p),\quad x_i=t, \qquad (3.37)$$

$$\widetilde{l}_{ij}(\theta)\leq\xi_{ij}(\theta)\leq\widetilde{u}_{ij}(\theta),\quad \theta+\tau_{ij}(\theta)\leq p,\ \theta\in T. \qquad (3.38)$$

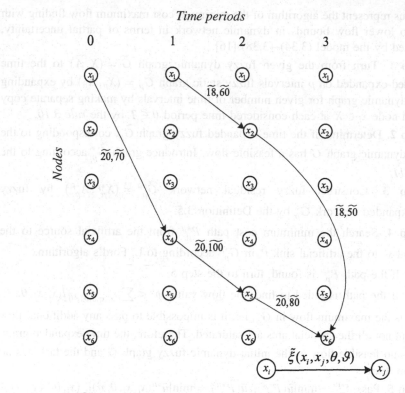

Fig. 3.53 Graph $\widetilde{G}_p$ with the flow of 38 units

Fig. 3.54 Given flow in the form of the fuzzy triangular number of (38,9,9) units

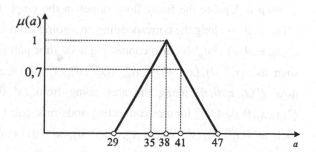

Fig. 3.55 The minimum transmission cost of (6980,647,729) flow units in the form of the fuzzy triangular number

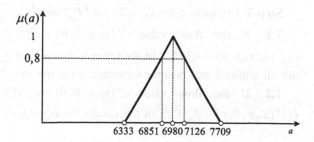

Let us represent the algorithm of the minimum cost maximum flow finding with nonzero lower flow bounds in dynamic network in terms of partial uncertainty, presented by the model (3.34)–(3.38) [16].

Step 1. Turn from the given fuzzy dynamic graph $\widetilde{G} = (X, \widetilde{A})$ to the time expanded-expanded on p intervals fuzzy static graph $\widetilde{G}_p = (X_p, \widetilde{A}_p)$ by expanding initial dynamic graph for given number of time intervals by making separate copy of each node $x_i \in X$ at each considered time period $\theta \in T$ by the *rule 3.10*.

Step 2. Determine, if the time-expanded fuzzy graph $\widetilde{G}_p$, corresponding to the initial dynamic graph $\widetilde{G}$ has a feasible flow. Introduce graph $\widetilde{G}_p^*$ according to the *rule 3.11*.

Step 3. Construct fuzzy residual network $\widetilde{G}_p^{*\mu} = (X_p^{*\mu}, \widetilde{A}_p^{*\mu})$ by fuzzy time-expanded network $\widetilde{G}_p^*$ by the Definition 3.5.

Step 4. Search the minimum cost path $\widetilde{P}_p^{*\mu}$ from the artificial source to the artificial s^* to the artificial sink t^* in $\widetilde{G}_p^{*\mu}$ according to L. Ford's algorithm.

4.1. If the path $\widetilde{P}_p^{*\mu}$ is found, turn to the step 5.

4.2. If the path is failed to find, the flow value $\tilde{\phi}^* < \sum_{\tilde{l}(x_i,x_j,\theta,\vartheta) \neq \tilde{0}} \tilde{l}(x_i, x_j, \theta, \vartheta)$, which is the maximum flow in $\widetilde{G}_p^*$, i.e. it is impossible to pass any additional flow unit, but not all the artificial arcs are saturated. Therefore, the time-expanded graph $\widetilde{G}_p$ has no feasible flow as the initial dynamic fuzzy graph $\widetilde{G}$ and the task has no solution. Exit.

Step 5. Pass $\tilde{\delta}_p^{*\mu} = \min[\tilde{u}(\widetilde{P}_p^{*\mu})], \tilde{u}(\widetilde{P}_p^{*\mu}) = \min[\tilde{u}^{*\mu}(x_i, x_j, \theta, \vartheta)], (x_i, \theta), (x_j, \vartheta) \in \widetilde{P}_p^{*\mu}$ flow units along the minimum cost path $\widetilde{P}_p^{*\mu}$.

Step 6. Update the fuzzy flow values in the graph $\widetilde{G}_p^*$: replace the fuzzy flow $\tilde{\xi}^*(x_j, x_i, \theta, \vartheta)$ along the corresponding arcs going from (x_j^*, θ) to (x_i^*, ϑ) from $\widetilde{G}_p^*$ by $\tilde{\xi}^*(x_j, x_i, \theta, \vartheta) - \tilde{\delta}_p^{*\mu}$ for arcs connecting node-time pair $(x_i^{*\mu}, \vartheta)$ with $(x_j^{*\mu}, \theta)$ in $\widetilde{G}_p^{*\mu}$, such as $((x_i^{*\mu}, \vartheta), (x_j^{*\mu}, \theta)) \in \widetilde{A}_p^*$, $((x_i^{*\mu}, \vartheta), (x_j^{*\mu}, \theta)) \in \widetilde{A}_p^{*\mu}$ and replace the fuzzy flow $\tilde{\xi}^*(x_i, x_j, \theta, \vartheta)$ along the arcs going from (x_i^*, θ) to (x_j^*, ϑ) from $\widetilde{G}_p^*$ by $\tilde{\xi}^*(x_i, x_j, \theta, \vartheta) + \tilde{\delta}_p^{*\mu}$ for arcs connecting node-time pair $(x_i^{*\mu}, \theta)$ with $(x_j^{*\mu}, \vartheta)$ in $\widetilde{G}_p^{*\mu}$, such as $((x_i^{*\mu}, \theta), (x_j^{*\mu}, \vartheta)) \in \widetilde{A}_p^*$, $((x_i^{*\mu}, \theta), (x_j^{*\mu}, \vartheta)) \in \widetilde{A}_p^{*\mu}$. Replace $\tilde{\xi}^*(x_i, x_j, \theta, \vartheta)$ by $\tilde{\xi}^*(x_i, x_j, \theta, \vartheta) + \tilde{\delta}_p^{*\mu} \widetilde{P}_p^{*\mu}$ and turn to the step 7.

Step 7. Compare $\tilde{\xi}^*(x_i, x_j, \theta, \vartheta) + \tilde{\delta}_p^{*\mu} \widetilde{P}_p^{*\mu}$ and $\sum_{\tilde{l}(x_i,x_j,\theta,\vartheta) \neq \tilde{0}} \tilde{l}(x_i, x_j, \theta, \vartheta)$:

7.1. If the flow value $\tilde{\xi}^*(x_i, x_j, \theta, \vartheta) + \tilde{\delta}_p^{*\mu} \widetilde{P}_p^{*\mu}$ of the minimum cost $\tilde{c}(\tilde{\xi}^*(x_i, x_j, \theta, \vartheta) + \tilde{\delta}_p^{*\mu} \widetilde{P}_p^{*\mu})$ of the value $\tilde{\sigma}^*$ is less than $\sum_{\tilde{l}(x_j,x_i,\vartheta,\theta) \neq \tilde{0}} \tilde{l}(x_i, x_j, \theta, \vartheta)$, i.e. not all artificial arcs become saturated, go to the step 3.

7.2. If the flow value $\tilde{\xi}^*(x_i, x_j, \theta, \vartheta) + \tilde{\delta}_p^{*\mu} \widetilde{P}_p^{*\mu}$ of the minimum cost $\tilde{c}(\tilde{\xi}^*(x_i, x_j, \theta, \vartheta) + \tilde{\delta}_p^{*\mu} \widetilde{P}_p^{*\mu})$ of the value $\tilde{\sigma}^*$ is equal to $\sum_{\tilde{l}(x_j,x_i,\vartheta,\theta) \neq \tilde{0}} \tilde{l}(x_i, x_j, \theta, \vartheta)$, i.e.

all arcs from the artificial source to the artificial sink become saturated, then the value $\tilde{\xi}^*(x_i, x_j, \theta, \vartheta) + \tilde{\delta}_p^{*\mu}\widetilde{P}_p^{*\mu}$ is required value of maximum flow $\tilde{\sigma}^*$ of the minimum cost $\tilde{c}(\tilde{\xi}^*(x_i, x_j, \theta, \vartheta) + \tilde{\delta}_p^{*\mu}\widetilde{P}_p^{*\mu})$. In this case the total flow along the artificial arcs connecting the node-time pairs $(t, \forall \theta \in T)$ with $(s, \forall \theta \in)T$, which is equal to $\sum_{\theta=0}^{p}\tilde{\xi}^*(t, s, \forall \theta \in T, \forall \theta \in T)$ in $\widetilde{G}_p^*$ determines the feasible flow in the time-expanded graph $\widetilde{G}_p$ with the flow value $\sum_{\theta=0}^{p}\tilde{\xi}^*(t, s, \forall \theta \in T, \forall \theta \in T) = \tilde{\sigma}$ of the minimum cost. Turn to the graph $\widetilde{G}_p$ from the graph $\widetilde{G}_p^*$ according to the *rule 3.5*. The network $\widetilde{G}_p(\tilde{\xi})$ is obtained. Turn to the step 8.

Step 8. Construct the residual network $\widetilde{G}_p^{\mu}(\tilde{\xi})$ taking into account the feasible flow vector $\tilde{\xi} = (\tilde{\xi}(x_i, x_j, \theta, \vartheta))$ in $\widetilde{G}_p(\tilde{\xi})$ adding the artificial source and sink and the arcs with infinite arc capacity, connecting s' with true sources and t' with true sinks according to the *rule 3.12*.

Step 9. Define the shortest path $\widetilde{P}_p^{\mu}$ according to the breadth-first-search from s' в t' in the constructed residual network $\widetilde{G}_p^{\mu}(\tilde{\xi})$.

9.1. Go to the step 10 if the augmenting path $\widetilde{P}_p^{\mu}$ is found.

9.2. If the path is failed to find, then the maximum flow $\tilde{\xi}(x_i, x_j, \theta, \vartheta) + \tilde{\delta}_p^{\mu}\widetilde{P}_p^{\mu} = \tilde{v}(p)$ of the minimum cost $\tilde{c}(\tilde{\xi}(x_i, x_j, \theta, \vartheta) + \tilde{\delta}_p^{\mu}\widetilde{P}_p^{\mu})$ is obtained in the time-expanded static fuzzy graph, turn to the step 12.

Step 10. Pass the flow value $\tilde{\delta}_p^{\mu} = \min[\tilde{u}(\widetilde{P}_p^{\mu})]$, $\tilde{u}(\widetilde{P}_p^{\mu}) = \min[\tilde{u}^{\mu}(x_i, x_j, \theta, \vartheta)]$, $(x_i, \theta), (x_j, \vartheta) \in \widetilde{P}_p^{\mu}$ along the found path.

Step 11. Update the flow values in the graph $\widetilde{G}_p(\tilde{\xi})$: replace the flow $\tilde{\xi}(x_j, x_i, \theta, \vartheta)$ by $\tilde{\xi}(x_j, x_i, \theta, \vartheta) - \tilde{\delta}_p^{\mu}$ along the corresponding arcs, going from (x_j, θ) to (x_i, ϑ) from $\widetilde{G}_p(\tilde{\xi})$ for arcs, connecting node-time pair (x_i^{μ}, ϑ) with (x_j^{μ}, θ) in $\widetilde{G}_p^{\mu}(\tilde{\xi})$, such as $((x_i^{\mu}, \vartheta), (x_j^{\mu}, \theta)) \notin \widetilde{A}_p, ((x_j^{\mu}, \vartheta), (x_j^{\mu}, \theta)) \in \widetilde{A}_p^{\mu}$, and replace the flow $\tilde{\xi}(x_i, x_j, \theta, \vartheta)$ by $\tilde{\xi}(x_i, x_j, \theta, \vartheta) + \tilde{\delta}_p^{\mu}$ along the corresponding arcs, going from (x_i, θ) to (x_j, ϑ) from $\widetilde{G}_p(\tilde{\xi})$ for arcs, connecting node-time pair (x_i^{μ}, θ) with (x_j^{μ}, ϑ) in $\widetilde{G}_p^{\mu}(\tilde{\xi})$, such as $((x_i^{\mu}, \theta), (x_j^{\mu}, \vartheta)) \in \widetilde{A}_p, ((x_i^{\mu}, \theta), (x_j^{\mu}, \vartheta)) \in \widetilde{A}_p^{\mu}$. Replace the flow value in $\widetilde{G}_p$: $\tilde{\xi}(x_i, x_j, \theta, \vartheta)$ by $\tilde{\xi}(x_i, x_j, \theta, \vartheta) + \tilde{\delta}_p^{\mu}\widetilde{P}_p^{\mu}$ and turn to the step 8, starting from the new flow value.

Step 12. If the flow is $\tilde{\xi}(x_i, x_j, \theta, \vartheta) + \tilde{\delta}_p^{\mu} \times \widetilde{P}_p^{\mu} = \tilde{\rho}(p)$ in the graph $\widetilde{G}_p(\tilde{\xi})$ of the minimum cost $\tilde{c}(\tilde{\xi}(x_i, x_j, \theta, \vartheta) + \tilde{\delta}_p^{\mu} \times \widetilde{P}_p^{\mu})$ from the artificial source s' to the artificial sink t', defined by the set of paths $\widetilde{P}_p^{\mu}$, turn to the initial dynamic graph $\widetilde{G}$ such a way: delete artificial nodes s', t' and arcs, connecting them with other nodes. Thus, the flow of the value $\tilde{\rho}(p)$ of the minimum cost is obtained in the initial dynamic graph, which is equivalent to the flow from the sources (the initial node expanded to the p intervals) to the sinks (terminal node expanded to the p intervals) in $\widetilde{G}_p$ after deleting artificial nodes and each path, connecting nodes (s, ς) and $(t, \psi = \varsigma + \tau_{st}(\varsigma)), \psi \in T$, with the flow $\tilde{\xi}(s, t, \theta, \varsigma)$ of the cost $\tilde{c}(\tilde{\xi}(s, t, \theta, \varsigma))$, coming along, corresponds to the flow $\tilde{\xi}_{st}(\theta)$ of the cost $\tilde{c}(\tilde{\xi}_{st}(\theta))$.

3.6 Summary

Present chapter presents proposed methods of the dynamic flow determining in fuzzy networks. In particular, new methods of the maximum, minimum cost flow finding in fuzzy dynamic networks with nonzero lower flow bounds are proposed and illustrated by numerical examples.

References

1. Chabini I, Abou-Zeid M (2003) The Minimum cost flow problem in capacitated dynamic networks. Annual Meeting of the Transportation Research Board, Washington D.C., pp 1–30
2. Bozhenyuk A, Rozenberg I, Starostina T (2006) Analiz i issledovaniye potokov i zhivuchesti v transportnykx setyakh pri nechetkikh dannykh. Nauchnyy Mir, Moskva
3. Bozhenyuk A, Rogushina E (2011) Method of maximum flow finding in fuzzy temporal network. European researcher. Multi Sci Periodical 5–1(7):582–583
4. Miller-Hooks E, Patterson SS (2004) On solving quickest time problems in time-dependent, dynamic network. J Math Model Algorithms 3:39–71
5. Bozhenyuk A, Gerasimenko E (2014) Flows finding in networks in fuzzy conditions. In: Kahraman C, Öztaysi B (eds) Supply chain management under fuzzines, studies in fuzziness and soft computing, vol 313, Part III, pp 269–291. Springer, Berlin. doi:10.1007/978-3-642-53939-8_12
6. Bozhenyuk A, Gerasimenko E, Rozenberg I (2012) Algorithm of maximum dynamic flow finding in a fuzzy transportation network. In: Proceedings of east west fuzzy colloquium 2012 19th fuzzy colloquium, pp 125–132, 5–7 Sept
7. Fonoberova M (2006) On determining the minimum cost flows in dynamic networks. Bul Acad Ştiinţe Repub Mold Mat 1:51–56
8. Fonoberova M, Lozovanu D (2004) The optimal flow in dynamic networks with nonlinear cost functions on edges. Bullet Acad Sci Moldova Math 3(46):10–16
9. Bozhenyuk AV, Gerasimenko EM, Rozenberg IN (2013) Minimum cost flow defining in fuzzy dynamic graph. Izvestiya SFedU. Eng Sci 5(130):149–154
10. Dijkstra EW (1959) A note on two problems in connexion with graphs. Numer Math 1:269–271
11. Edmonds J, Karp RM (1972) Theoretical Improvements in algorithmic efficiency for network flow problems. J Assoc Comput Mach 19(2):248–264
12. Tomizawa N (1971) On some techniques useful for solution of transportation network problems. Networks 1:173–194
13. Gerasimenko EM (2014) Potentials method for minimum cost flow defining in fuzzy dynamic graph. Izvestiya SFedU. Eng Sci 4:83–89
14. Pouly A (2010) Minimum cost flows. 23 Nov 2010. http://perso.ens-lyon.fr/eric.thierry/Graphes2010/amaury-pouly.pdf

15. Bozhenyuk A, Gerasimenko E (2013) Algorithm for monitoring minimum cost in fuzzy dynamic networks. Inf Technol Manag Sci (RTU Press, Riga) 16(1):53–59. doi:10.2478/itms-2013-0008
16. Bozhenyuk A, Gerasimenko E, Rozenberg I, Perfilieva I (2015) Method for the minimum cost maximum flow determining in fuzzy dynamic network with nonzero flow bounds. In: Alonso José M, Bustince H, Reformat M (eds) Proceedings of the 2015 conference of the international fuzzy systems association and the european society for fuzzy logic and technology, vol 89, pp 385–392. 30th June–3rd July. Atlantic Press, Gijón, Asturias (Spain). ISBN (on-line) 978-94-62520-77-6. ISSN 1951-6851. doi:10.2991/ifsa-eusflat-15.2015.56

Chapter 4
Implementing the Program for Solving the Flow Tasks in Networks in Fuzzy Conditions

Practical implementation of the proposed flow algorithms in fuzzy conditions is considered with solving the certain logistics tasks and tasks of analysis on the network of Russian Federation, that reflected in the proposed software complex that realizes solving of proposed flow algorithms.

4.1 Functional Purpose of Proposed Program Complex

Program realizes the following algorithms:

1. Finding the maximum flow in the network with fizzy arc capacities.
2. Finding the maximum flow in the network with upper and lower flow bounds, presented in the fuzzy form.

Practical realization of the proposed algorithms is performed based on railway map of Russian Federation. GIS ObjectLand 2.6 is used as the source of the input data. IDE Microsoft Visual Studio Express 2012 for Windows Desktop is applied as the platform of realization. C# is the realization language.

Already existed tables from GIS ObjectLand describing the railway network of the Russian Federation are used as the main source of the input data.

The opportunity to use different ways of fuzzy numbers setting is either necessary condition, for example:

1. Triangular fuzzy numbers.
2. Trapezoidal fuzzy numbers.
3. Interval fuzzy numbers.

Fragment of CSV-file describing the region of the railway network of Russian Federation is presented in the Table 4.1.

© Springer International Publishing Switzerland 2017
A.V. Bozhenyuk et al., *Flows in Networks Under Fuzzy Conditions*,
Studies in Fuzziness and Soft Computing 346, DOI 10.1007/978-3-319-41618-2_4

Table 4.1 Example of the input data

ID of initial station	ID of terminal station	Upper flow bound	Lower flow bound
3844	3845	8	0
3836	3837	7	2

Accordingly the fields «ID of initial station» and «ID of terminal station» will be modified in initial and terminal nodes of the arc and the fields «Upper flow bound» and «Lower flow bound»—in upper and lower flow bounds.

Thus, any CSV-file of the given structure can be used as input data. In this case the names of specific fields can be changed by using the menu item "Settings".

Input data for the program are:

1. Calculated flow value in the fuzzy form.
2. Flow distribution along the network in the form of the list of paths and the flow along them.

Incremental report on the work of analysis algorithms.

The main screen form of the program is presented in the Fig. 4.1. Basic menu contains the following items:

1. File—allows the user to select the CSV-file for the data import.
2. Algorithm—allows the user to select between different flow algorithms.
3. Settings—causes the modal window «Application Settings», where you can view and edit all the necessary parameters.

The main form allows user to look through and edit uploaded data, view detailed report on the chosen algorithm.

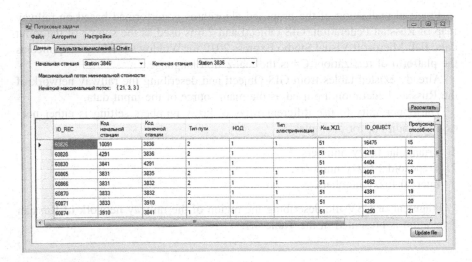

Fig. 4.1 The main user form of the program

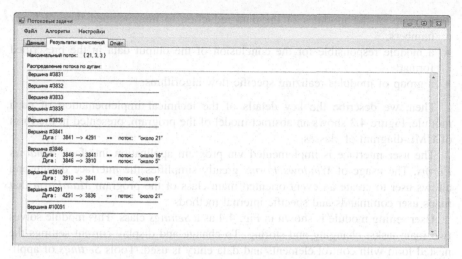

Fig. 4.2 The main user form in the playback mode

User has the opportunity to set initial and terminal points of the network on the tab «Data», which have important value in calculations on algorithm in the tab «Algorithms».

Example of viewing the output data of the algorithm for the maximum flow finding is presented in the Fig. 4.2. Key points of calculations performing with explanations and values of key local variables are presented «Incremental Report» field.

If it is necessary, the user can apply the «Settings» menu item and select the desired type of fuzzy numbers, which will be used in the calculations. One can change either the rules of the data extraction from downloadable CSV-file, particularly, to specify the name of the fields, which it is necessary to retrieve certain information from in the dialog menu «Settings».

4.2 Description of the Logical Structure of the Software Mode

Conventionally, the whole range of problems solved by the program can be divided into the following modules:

- user interface in a form of the control and display elements;
- editing and user preferences storing module (names of the fields, basic values, used types of fuzzy numbers, etc.);
- data import module from CSV-files with the conversion of the data into a form convenient for further processing;
- a group of generic classes on work with fuzzy numbers of different types;

- a group of the modules of the data reduction to the specified type of fuzzy numbers;
- a module responsible for the conclusion of the output data in the prescribed format;
- a group of modules realizing specific flow algorithms.

Then, we describe the key details of the technical implementation of each module. Figure 4.3 shows an abstract model of the program, presented in the form of UML-diagram of classes.

The user interface is implemented via program application interface *Windows Forms*. The usage of *Windows Forms* greatly simplifies the interface design and allows user to create an event-oriented main class of the program *MainForm* that maps user commands and specific internal methods.

User setting module is shown in Fig. 4.4 as a *Settings* class. This module solves two main tasks: changing and storing. To change and display current settings the nested form with control elements and data entry is used. Tools *Settings* of application *Windows Forms* that allows user to set default values and save the settings entered by the user throughout the entire cycle of the program is used for storing these settings.

Module of data import from CSV-files is presented in the form of classes *CSVImporter CSVDataTransformer* (Fig. 4.3). The tasks of the class *CSVImporter* are loading, reading user specified file, saving any changes and data conversion to the C# language types. The main purpose of *CSVDataTransformer* is creation of

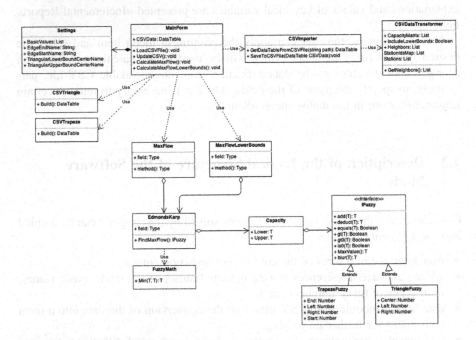

Fig. 4.3 The class diagram of the program

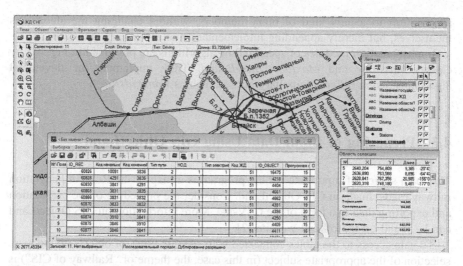

Fig. 4.4 The sample data based on the objects selection of the type Driving

the language structures, which will contain the information required by modules that implement proposed flow algorithms. Thus, for example, it is necessary to have the so-called list of neighbors (pairs of the form {name of the vertice}: [array of adjacent nodes]) and create a matrix of arc capacities for correctness of J. Edmonds and R. Karp's algorithm. At this point we come to the presentation of network in the form of a directed graph with fuzzy parameters.

Since the program should operate different types of fuzzy numbers the majority of classes is implemented using generic classes. Universal classes encapsulate operations that are not related to any particular type of data [1]. The different types of fuzzy numbers (triangular, trapezoidal, etc.) will be used as a specific data type. Classes of fuzzy numbers *TriangleFuzzy* and *TrapezeFuzzy* implement universal interface *IFuzzy*, which specifies the whole necessary further set of operations on fuzzy numbers.

For any arc of the graph it is necessary to specify the value of the fuzzy arc capacity and either upper or lower flow bounds can be specified for some arcs. Generic class *Capacity* that contains the properties *Upper* and *Lower* for upper and lower bounds of streams is created for it. The instances of classes that implement the interface *IFuzzy* (particularly, *TriangleFuzzy* and *TrapezeFuzzy*) are used as the values of flow bounds.

Then the group of modules that performs the conversion capacities of the arc to the specified type of fuzzy numbers follows. Figure 4.3 shows two modules of this group: *CSVTriangle* and *CSVTrapeze*. These modules create instances of the class *Capacity* with appropriate types of fuzzy numbers. After this step, it is possible to implement all algorithms in the form of generic classes and methods that do not depend on the choice of a particular type of fuzzy numbers that greatly simplifies the implementation and makes the code more flexible. So, if necessary, one can always turn to a new kind of fuzzy numbers, without changing the algorithms themselves, but merely appending implementation of the *IFuzzy.* interface.

Implementation modules of specific flow algorithms are presented in Fig. 4.3 by classes *MaxFlow* and *MaxFlowLowerBound.* They implement the developed algorithms, using different auxiliary classes such as *EdmondsKarp.*

4.3   Preparing of Input Data Using GIS ObjectLand

The main source of input data for the program is a map of the Russian Federation railway roads, made in GIS ObjectLand. Consider the procedure of the preparation and exporting data in a form of the CSV-file. More information on GIS ObjectLand can be found on the official website [2].

To gain access to the tables connected with specific types of objects on the map, selection of the appropriate subject (in this case, the theme of "Railway of CIS") is implemented. It is possible to get a sample of related data only for selected objects of the same type, so it is necessary to allow selection for the type Drivings before using the selection tool in the topic legend. Figure 4.4 shows the result of the described above steps for sampling of some of the Rostov railway node.

Further, use the command Selection > Export > in text format > delimited. After that it will be asked to specify a file, which you want to export data to (CSV-file will be created when choosing this method of export). It is necessary to enable the option "Export field names."

4.4   Summary

This chapter contains the description of program realization of the proposed flow methods in static and dynamic networks in fuzzy conditions via GIS «ObjectLand». Practical realization of the proposed algorithms is performed based on railway map of Russian Federation. GIS ObjectLand 2.6 is used as the source of the input data. Realized program operates different types of fuzzy numbers and calculates the maximum flow with zero and nonzero lower flow bounds.

References

1. Generic classes (C# Programming Guide). https://msdn.microsoft.com/en-us/library/sz6zd40f.
   aspx
2. ObjectLand Geoinformation system. http://www.objectland.ru/

Conclusion

This monograph is devoted to researches in flow tasks in networks in fuzzy conditions. Approaches to maximum flow finding, the minimum cost flow finding in static and dynamic networks in terms of fuzziness and partial uncertainty are supposed. Decisions of these tasks allow to find the maximum cargo traffic between selected nodes on the road, identify the paths of transferring cargo of the minimum cost map and solve these tasks taking into account limited time horizon.

Extreme flow tasks in fuzzy conditions nowadays are poorly studied. Despite of the researches of foreign authors in the field of flows, there are unexplored issues related to flows finding, taking into account fuzzy nonzero lower flow bounds, arc capacities, transmission costs, crisp time parameters, depending on the flow departure times. The method of operating with fuzzy flows in the form of fuzzy numbers is proposed. This method uses the centers of fuzzy numbers and blurs the result at the final step of the algorithm. Nonstandard operation of subtraction is implemented in the method, as it doesn't lead to the strong blurring of the resulting number and negative flow values.

The new scientific results described in the present book are constructing of new mathematical models of extreme flow tasks in fuzzy networks and modification of the existed methods of the flow tasks solving. In particular, the following tasks are proposed in the monograph and illustrated by the numerical examples:

1. Methods of the maximum flow finding in the network with zero and nonzero lower flow bounds. These methods allow, despite of the existed, find the maximum flow with zero and nonzero lower flow bounds set in the fuzzy form.
2. Methods of the minimum cost flow finding in networks, which parameters are zero and nonzero upper, lower flow bounds and transmission costs that allow find the minimum cost flow with fuzzy upper, lower flow bounds and transmission costs.
3. Methods of the maximum flow finding in dynamic network with fuzzy zero and nonzero lower and upper flow bounds that differ from analogues that allow to take into account dependence of fuzzy lower and upper flow bounds from departure time.

© Springer International Publishing Switzerland 2017
A.V. Bozhenyuk et al., *Flows in Networks Under Fuzzy Conditions*,
Studies in Fuzziness and Soft Computing 346, DOI 10.1007/978-3-319-41618-2

4. Methods of the minimum cost flow finding in dynamic networks with fuzzy arc capacities and costs that differ from analogues that allow take into account dependence of fuzzy arc capacities and transmission costs from departure time.
5. Methods of the minimum cost flow finding in dynamic networks with fuzzy nonzero lower flow bounds and costs that differ from analogues that allow take into account dependence of fuzzy upper, lower flow bounds and transmission costs from departure time.
6. Program module implementing while finding fuzzy maximum flow with zero and nonzero lower flow bounds used with GIS ObjectLand.

Our future researchers lie in the field of flows finding in dynamic networks with given vitality degree and increasing of vitality degree in fuzzy networks.

Printed in the United States
By Bookmasters